# SpringerBriefs in Electrical and Computer Engineering

More information about this series at http://www.springer.com/series/10059

Omid Ardakanian · S. Keshav
Catherine Rosenberg

# Integration of Renewable Generation and Elastic Loads into Distribution Grids

Springer

Omid Ardakanian
University of California, Berkeley
Berkeley, CA
USA

Catherine Rosenberg
University of Waterloo
Waterloo, ON
Canada

S. Keshav
University of Waterloo
Waterloo, ON
Canada

ISSN 2191-8112          ISSN 2191-8120   (electronic)
SpringerBriefs in Electrical and Computer Engineering
ISBN 978-3-319-39983-6          ISBN 978-3-319-39984-3   (eBook)
DOI 10.1007/978-3-319-39984-3

Library of Congress Control Number: 2016941086

Printed on acid-free paper

This Springer imprint is published by Springer Nature
The registered company is Springer International Publishing AG Switzerland

# Preface

Variable-power distributed energy resources, such as solar photovoltaic and storage systems, and high-power elastic loads, such as electric vehicle chargers, are being installed at a phenomenal rate in power distribution systems. Such active end-nodes can affect the reliable operation of the grid if they are not controlled properly. Yet there is no consensus among various stakeholders in the power industry on the significance and impacts of these cutting-edge technologies. This brief focuses on the challenges of integrating active end-nodes into low-voltage distribution grids and the potential for pervasive measurement and control to address these challenges. A mathematical framework is presented for the joint control of active end-nodes at scale, and it is shown through extensive numerical simulations that proper control of active end-nodes can significantly enhance reliable and economical operation of the power grid.

Berkeley, USA
Waterloo, Canada
Waterloo, Canada

Omid Ardakanian
S. Keshav
Catherine Rosenberg

# Acknowledgements

The authors would like to acknowledge the valuable suggestions made by Claudio Canizares, Jeffrey Taft, Alexandra von Meier, Ansis Rosmanis, and Come Carquex.

# Contents

# Symbols

| | |
|---|---|
| $\alpha_i^c$ | Rated charge capacity of storage $i$ |
| $\alpha_i^d$ | Rated discharge capacity of storage $i$ |
| $\beta_i$ | Rated capacity of EV charger $i$ |
| $\eta_i^c$ | Charge efficiency of storage $i$ |
| $\eta_i^d$ | Discharge efficiency of storage $i$ |
| $\gamma_i^c$ | Charge efficiency of EV battery $i$ |
| $\mathbf{A}^b$ | Bus association matrix for storage systems |
| $\mathbf{A}^e$ | Bus association matrix for EV chargers |
| $\mathbf{A}^l$ | Bus association matrix for inelastic loads |
| $\mathbf{A}^s$ | Bus association matrix for PV systems |
| $\mathcal{B}$ | Set of buses |
| $\mathcal{B}_{\mathcal{Z}}$ | Set of buses that represent balancing zones |
| $\mathcal{B}_i$ | Set of buses downstream of bus $i$ |
| $\mathcal{E}$ | Set of EV chargers |
| $\mathcal{I}$ | Set of inelastic loads |
| $\mathcal{J}$ | Set of PV systems |
| $\mathcal{L}$ | Set of primary distribution lines |
| $\mathcal{L}^i$ | Set of lines located on the unique path from the substation to bus $i$ |
| $\mathcal{S}$ | Set of battery storage systems |
| $\mathcal{T}$ | Set of time slots |
| $\bar{c}_i$ | Maximum SOC of storage $i$ |
| $\bar{p}_i^b(t)$ | Maximum acceptable discharge power of storage $i$ in time slot $t$ |
| $\bar{p}_i^e(t)$ | Maximum acceptable charge power of EV charger $i$ in time slot $t$ |
| $\bar{p}_i^s(t)$ | Available real power at PV system $i$ in time slot $t$ |
| $\bar{s}_i^s$ | Apparent power rating of inverter $i$ |
| $\tau$ | Length of a time slot |
| $\underline{c}_i$ | Minimum SOC of storage $i$ |
| $\underline{p}_i^b(t)$ | Maximum acceptable charge power of storage $i$ in time slot $t$ |
| $\underline{p}_i^e(t)$ | Minimum acceptable charge power of EV charger $i$ in time slot $t$ |

$\xi_{ij}$          Setpoint associated with a line or a transformer located between bus $i$ and $j$

$c_i(t)$         SOC of storage $i$ in time slot $t$

$d_i$          Charging deadline of EV charger $i$

$e_i(t)$         Energy required to fulfill charging demand of EV $i$ in time slot $t$

$p_i^b(t)$        Real power contribution of storage $i$ in time slot $t$

$p_i^e(t)$        Charge power of EV charger $i$ in time slot $t$

$p_i^l(t)$        Real power consumption of inelastic load $i$ in time slot $t$

$p_i^s(t)$        Real power contribution of solar inverter $i$ in time slot $t$

$p_i(t)$         Total real power consumed at bus $i$ in time slot $t$

$P_{ij}(t)$       Real power flow from bus $i$ to downstream bus $j$ in time slot $t$

$q_j^c(t)$        Reactive power provided by capacitors at bus $j$ in a given time slot

$q_i^l(t)$        Reactive power consumption of inelastic load $i$ in time slot $t$

$q_i^s(t)$        Reactive power contribution of solar inverter $i$ in time slot $t$

$q_i(t)$         Total reactive power consumed at bus $i$ in time slot $t$

$Q_{ij}(t)$       Reactive power flow from bus $i$ to downstream bus $j$ in time slot $t$

$r_{ij}$          Resistance of line connecting bus $i$ to bus $j$

$S_{ij}(t)$        Apparent power flow from bus $i$ to downstream bus $j$ in time slot $t$

$v_0$          Voltage magnitude at the substation bus

$v_{max}$       Upper voltage limit in the distribution network

$v_{min}$        Lower voltage limit in the distribution network

$v_i(t)$         Voltage magnitude at bus $i$ in time slot $t$

$x_{ij}$          Reactance of line connecting bus $i$ to bus $j$

$z_{ij}$          Impedance of line connecting bus $i$ to bus $j$

Note that the upright boldface letters represent matrices.

# Acronyms

| | |
|---|---|
| AC | Alternating current |
| AIMD | Additive-increase multiplicative-decrease algorithm |
| AMI | Advanced metering infrastructure |
| BMS | Battery management system |
| CSP | Charge service provider |
| DC | Direct current |
| DER | Distributed energy resources |
| DOD | Depth of discharge |
| DOPF | Distribution optimal power flow |
| DR | Demand response |
| DSO | Distribution system operator |
| EV | Electric vehicle |
| EVSE | Electric vehicle supply equipment |
| GHG | Greenhouse gas |
| ICT | Information and communications technology |
| LP | Linear programming |
| LPWAN | Low-power wide area networks |
| LTC | Transformer load tap changer |
| MIP | Mixed integer programming |
| MPC | Model predictive control |
| NUM | Network utility maximization |
| OPF | Optimal power flow |
| PEV | Plug-in electric vehicle |
| PV | Photovoltaics |
| SAE | Society of Automotive Engineers |
| SOC | Battery state of charge |
| TCL | Thermostatically controlled load |
| TCP | Transmission control protocol |
| V2G | Vehicle-to-grid |
| VAR | Volt-ampere reactive |

# Chapter 1
# Introduction

**Abstract** Large-scale integration of variable-power distributed energy resources (DER), such as solar photovoltaics and storage systems, and high-power elastic loads, such as electric vehicle chargers, into low-voltage distribution grids can pose serious challenges to power system operators. This chapter discusses how pervasive measurement and control can be used to address these challenges and enhance the reliable and economical operation of the grid. It also specifies design goals for future grid control mechanisms, and proposes a new approach to the control of DER and elastic loads.

## 1.1 Traditional Grid

The North American power grid is one of the largest machines ever built. This gigantic, carbon-intensive legacy system comprises thousands of power stations producing electricity to serve demands of millions of geographically dispersed electrical loads, and has an enormous number of transmission and distribution lines and transformers connecting the power stations to distribution substations and downstream loads. Despite the scale and complexity of the power grid, its fundamental task is surprisingly simple: it delivers power to loads while ensuring *reliability*[1] and low cost. From the early days of the grid, reliability has always been of utmost importance and this perspective has been reflected in its planning and operation. In particular, electric utilities size and operate the grid in a way that the available generation capacity almost always[2] exceeds demand peaks and the transmission and distribution capacity is almost always sufficient to deliver power to the loads. The success of this approach is reflected in the fact that today customers in many parts of the world take it for

---

[1]Power system reliability generally describes the continuity of electric service to customers with a voltage and a frequency within prescribed ranges.

[2]A widely accepted benchmark value for reliability in the United States is the "one-day-in-ten-years criterion", which means that the system-wide generation capacity is expected to fall short of demand once every ten years [21].

© The Author(s) 2016
O. Ardakanian et al., *Integration of Renewable Generation and Elastic Loads into Distribution Grids*, SpringerBriefs in Electrical and Computer Engineering,
DOI 10.1007/978-3-319-39984-3_1

granted that lights turn on as they flip a switch. They do not even notice that power system operators are taking measures to constantly and precisely balance supply and demand.

The traditional power grid has the following characteristics that are relevant to the discussion of distribution grid control, the focus of this chapter:

- **Generation**—Power stations are typically centralized and dispatchable, i.e., their power output can be adjusted at the request of system operators, though some are more responsive than others. In many countries, most power stations burn fossil fuels to produce electricity, contributing to carbon emissions. Power stations are interconnected by high voltage transmission lines forming a mesh network with many redundant pathways. Hence, there is a clear physical and structural separation between generation resources and loads, which are typically connected to distribution feeders.
- **Loads**—Residential and commercial loads are mostly *inelastic*, i.e., their demand cannot be controlled or shaped. Although it is difficult to accurately predict the demand of a single load at a given time, the aggregate demand of a large number of loads across the grid behaves in a relatively predictable manner. This enables the system operators to schedule generation units a day or an hour in advance.
- **Customers**—In the traditional power grid, customers are information poor, control poor, yet energy rich. That is, they do not receive real-time electricity price or other signals that indicate the state of the power system, they have no means to control or schedule their loads, yet they are permitted to consume electricity at will as long as their demand is lower than a limit enforced by a circuit breaker.
- **Storage**—Physical energy storage is expensive and scarce. Thus, electricity must be produced and consumed instantaneously.[3]
- **Distribution networks**—Unlike transmission networks, legacy distribution networks are equipped with little instrumentation beyond the substation for cost reasons. Hence, distribution system operators (DSOs) have no way of determining the state of the network and cannot initiate remote remedial actions. Even the location of an outage in the distribution network is often determined by customer calls, unless it affects a manned substation [21]. Given that the grid is over-provisioned by design and traditional distribution networks are mostly radial with unidirectional power flow, service reliability is not at risk, despite having poorly monitored and controlled circuits.

---

[3] A small amount of energy storage in the form of rotational inertia is implicit in the traditional grid. This helps the operators to balance load and generation within a short time scale.

The consequence of these characteristics is that uncertainties are minimal and manageable in traditional power systems. This is because most generation units are dispatchable, and the overall demand does not vary drastically over a short period of time and can be predicted with sufficient accuracy several hours in advance.

However, the century-old grid is extremely under-utilized and inefficient; it is sized to meet the peak demand, which tends to occur only a few hours a year. This design principle is essential to preserve reliability when demand elasticity and storage capacity are very limited, but leads to a large carbon footprint.

## 1.2 Drivers of Change

In recent years, the traditional grid has undergone substantial changes due to the integration of several demand-side technologies into low-voltage distribution networks. This section introduces these low-carbon technologies and their potential impact on the grid, highlighting the growing need for control in distribution systems. A more comprehensive impact study is presented in Chap. 2. Our focus is restricted to the three most important technologies in distribution systems, namely renewable energy systems such as solar photovoltaics (PV) and wind turbines, electric vehicles (EVs),[4] and battery storage systems.

### 1.2.1 Renewable Energy Systems

Renewable generation costs have declined substantially in many parts of the world mainly due to sustained technology progress and improved financing conditions. For example, solar power has reached grid parity[5] in several jurisdictions today and is expected to soon become competitive with retail electricity in many other jurisdictions, even if existing investment tax credits expire [9]. This has led to increased deployment of rooftop solar panels in residential and commercial sectors, making solar PV distributed generation one of the fastest-growing renewable generation technologies at the present time.

Unfortunately, a high concentration of inherently-variable solar generation (and other types of renewable generation) in distribution networks is a mixed

---

[4]This work focuses on plug-in electric vehicles (PEVs), which are a subset of EVs that can be charged from the grid. But, for convenience, these two terms are often used interchangeably.

[5]Grid parity occurs when the levelized cost of solar PV (over a 20–25 year horizon) becomes less than or equal to the retail electricity price.

blessing. First, increased uncertainty in generation capacity both complicates generation planning [23] and increases the need for frequency regulation by fast-ramping fossil fuel power plants, which can actually *increase* overall carbon emissions.[6] Second, solar PV generation can surpass the feeder loading in some periods, resulting in *reverse power flow* and voltage rise toward the end of the feeder [15]. Reverse flows can cause protection coordination problems and the overuse of voltage regulating devices and circuit breakers, shortening their expected life cycle. Third, curtailing inexpensive solar power, i.e., accepting less solar power than what is available and displacing it by higher-priced resources, might be necessary to avoid distribution network problems in some situations [18]. However, in many jurisdictions, electric utilities need to pay for solar generation even if it is curtailed. This leads to the paradox of a large installed base of solar generation with small actual usage of solar power, yet with higher electricity bills for all.

Growing concerns over the impacts of distributed renewable generation on power system planning and operation have led to the design of sophisticated inverters that are capable of on-demand curtailment of real power and reactive power adjustment in addition to their basic task of converting direct current output of renewable energy systems to alternating current [16]. These *smart inverters* can be controlled to tackle overvoltage and unbalance conditions and prevent reverse flow [12, 27, 32]. Hence, a measurement, communication, and control infrastructure is essential for taking full advantage of the smart inverters.

## 1.2.2 Electric Vehicles

The transportation sector is by far the largest consumer of petroleum, and the second largest contributor to global greenhouse gas (GHG) emissions, accounting for about 23 % of the global GHG emissions in 2012 [14]. Transportation electrification could alleviate growing concerns over climate change and petroleum scarcity. Therefore, many governments have issued mandates to incentivize the adoption of electric vehicles so as to reduce their reliance on petroleum and cut down GHG emissions.

The EV market is growing fast. Global EV stock exceeded 665,000 in 2014, which is about 0.08 % of the total passenger car stock at present [10], and it is anticipated that EVs will account for 64 % of U.S. light-vehicle sales and will comprise 24 % of the U.S. light-vehicle fleet by 2030 [4]. Several automakers, including Nissan, Chevrolet, Toyota, General Motors, Ford, Honda, Audi, BMW, Renault, BYD, and Tesla, have embraced this technological shift and have released all-electric and plug-in hybrid EV models for the mass-market.

However, widespread EV adoption poses several new challenges for electric utilities and distribution system operators. At moderate to high penetration levels, uncontrolled electric vehicle charging can increase the peak load and energy losses, overload or *congest* distribution lines and transformers, and cause voltage swings and

---

[6]Germany has already encountered this problem, known as the *Energiewende* paradox [11].

phase imbalance in the distribution system [7, 20, 26]. Unrelieved congestion can overheat transformer windings and accelerate degradation of line and transformer insulation, leading to premature equipment failure. Excessive voltage drop can cause damage to electrical appliances.

Even at low penetration levels, there are likely to be certain neighbourhoods with high penetration levels [8]. For instance, the state of California's share of total nationwide plug-in EV registrations reached 45 % in 2014, accounting for 129,470 units out of the 286,842 PEVs registered in the U.S. since 2010 [5]. Uncoordinated EV charging could have detrimental impacts on the distribution network in these eco-friendly and eco-trendy neighbourhoods, even if the EV penetration level is relatively low in the entire distribution network.

To accommodate the EV charging load, utilities can take either of two approaches. The first approach is to make the required investment to upgrade distribution circuits as they become overloaded. The second approach is to exploit the elasticity of the EV charging load and a broadband communication network overlaid on the distribution network to directly control *smart* EV chargers.[7] The second approach significantly reduces the required reinforcement investment to accommodate higher EV penetration levels [25], assuming that the required measurement, communication, and control infrastructure is already in place.

## 1.2.3 Battery Storage Systems

With the growing interest in battery storage systems, especially when paired with solar PV installations, and the announcement of Tesla's Gigafactory, the world's largest lithium-ion battery factory, the cost per kilowatt-hour of battery storage systems is expected to fall dramatically by 2020 [30].[8] This will increase the number of battery storage systems connected to distribution feeders, as well as those integrated into the transmission network.

Battery storage systems offer several benefits to many aspects of the grid. For example, storage can be used to shave peaks and level loads, reducing carbon emissions, and transmission and distribution losses. It can also help operators better match supply with demand to maintain frequency. Indeed, the charge and discharge powers of storage systems can be adjusted even faster than the operating setpoint of fast-ramping generators that provide regulation service, making them excellent alternatives for balancing the future grid [6]. As a third example, storage can reduce the curtailment of renewable energy, which is necessary when there is a risk of

---

[7]Smart EV chargers choose a charging power/rate based on control signals that they receive from the grid. They are capable of charging EVs at any rate below the maximum charge power that they support.

[8]The cost per kilowatt-hour of battery packs used by market-leading EV manufacturers was approximately US$300 in 2014 [24].

over-generation or the network access link from a solar or a wind farm is overly congested and therefore cannot transmit excess power to other locations.

It should be clear that the careful control of storage can reduce reverse power flow, the need for frequency regulation from the grid, wasteful and expensive renewable generation curtailment, and overall carbon emissions. Whether storage systems actually offer any of these benefits depends on how they are owned and operated in practice. For example, a control strategy that tries to minimize solar curtailment would charge storage only from solar panels and not from the grid to ensure that storage capacity is available when the sun is shining, whereas a control strategy that provides frequency regulation services would keep storage roughly half-full at all times to support both up and down regulation. Thus, choosing the correct storage operation strategy is a complex problem that we consider later in this brief.

### 1.2.4 Emerging Challenges and Opportunities

The three demand-side technologies presented in the prior section can be classified into two major types. The first type introduces uncertainties in generation and load at various time scales. These uncertainties threaten the overall reliability of the grid and mitigating them is quite costly, requiring additional operating reserves. Renewable generation technologies are examples of this type. The second type provides additional control flexibility to the operators, thereby enabling operators to quickly react to operating conditions. Electric vehicle chargers, smart renewable power inverters, and storage systems, collectively referred to henceforth as *active end-nodes*, are examples of this type. The synergy between these two types of technology could enhance system reliability if they are carefully controlled by the grid; otherwise, these technologies impose new challenges to grid operators and can impair reliability. We now consider this synergy in more detail.

## 1.3 Enabling Technologies for the Control of Active End-Nodes

Addressing the challenges posed by the integration of the disruptive load and generation technologies discussed in Sect. 1.2 requires the sophisticated control of active end nodes, which is the focus of our work. The introduction of these controls will morph the traditional grid into an intelligent, more reliable and economical, and less carbon-intensive network, referred to as the "smart grid".

### 1.3.1 Pervasive Measurement and Communication

The smart grid heavily relies on the availability of pervasive measurement, communications, and computation in distribution networks to support two-way flow of information between the grid and its customers. The possibility of receiving near real-time information enables seamless control of active end-nodes at scale to ensure reliability and efficiency. Hence, pervasive measurement and communication, especially in the last mile of distribution networks, are the key enabling technologies for preventing loss of grid reliability due to the widespread adoption of the active end-nodes. To this end, several utilities in the United States have begun to install relatively inexpensive, high-precision phasor measurement units, called *micro-synchrophasors*, to monitor their distribution circuits. A micro-synchrophasor device provides high-sample-rate synchronized voltage and current magnitude and angle measurements; these measurements can be used in various diagnostic and control applications [22]. It is anticipated that many more distribution networks will soon be equipped with such measurement devices [31].

In addition to the synchrophasor technology, millions of *smart meters* have been rolled out around the world in recent years to collect more frequent electricity consumption data from customers and, in return, receive price and other signals from the grid. The two-way communication between meters and the grid can be used to shave demand peaks through time-of-use pricing and demand response (DR) programs.[9] Pervasive communication is possible using either existing cellular networks or new low-power wide area networks (LPWAN) [33].

### 1.3.2 Pervasive Control

Smart grid operators have to deal with fast-timescale dynamics that were absent in the traditional grid. These dynamics are introduced by fluctuating supply and demand and are observed even at low penetrations of active end-nodes. Thus, fast-timescale control is necessary to counteract these fluctuations, averting reliability and power quality problems. This is made possible by the deployment of pervasive control elements, in the form of embedded processors, that can be co-located with the element to be controlled. For example, pervasive control allows us to control EV charging rates at a fine timescale, on the order of seconds.

Smart grid customers can receive price and control signals from the grid and will be capable of setting and enforcing preferences and deadlines for their elastic loads.

---

[9]Note that the advanced metering infrastructure (AMI) deployed in some jurisdictions for billing purposes operate at at time scale of 10 min to an hour, and therefore are unable to support applications that require frequent communications between the electric utility and customers, such as demand response.

*Elastic loads* are defined as a class of loads that can be controlled within a limited range. Examples include dedicated storage, electric vehicles, and thermostatically controlled loads (TCLs) with inherent thermal energy storage. Depending on the jurisdiction, elastic loads might be controlled directly by electric utilities or the customers who own these loads. In the latter case, signals issued by the utility along with customers' input can be incorporated into the control process.

## 1.4  Need for a New Approach for Control of Active End-Nodes

Traditionally, control has focused on generation, since loads are viewed as being uncontrollable. Cost-effectively scheduling dispatchable generation units to meet forecasted load and reserve requirements involves solving security-constrained *unit commitment*[10] and security-constrained *economic dispatch* optimization problems. Unit commitment and economic dispatch are performed in day-ahead and real-time electricity markets, respectively [34].[11] Both problems incorporate a set of compli-cated constraints, including generating unit and transmission network constraints, and are cast as optimal power flow (OPF) problems [1]. However, these optimization problems do not include numerous distribution network constraints; this is because distribution networks are typically over-provisioned and unlikely to be stressed by a specific dispatch decision. Moreover, it is quite difficult to incorporate end-node objectives in the objective functions of these problems since they might be competing with the objectives of grid operators. Thus, traditional grid operation cannot be easily extended to control solar PV inverters, EV chargers, and storage systems which are connected to distribution networks [28], although this is necessary to ensure that distribution network constraints are not violated.

At the same time, using ad hoc controls in the distribution network can make the distribution control system unsustainable and insecure, potentially leading to chaotic situations [29]. Hence, new mechanisms are required to control the active end-nodes at scale in the distribution network. These mechanisms should be developed as extensions of the mechanisms that are already in place for balancing the grid.

---

[10]Unit commitment is a mixed integer programming (MIP) problem with many variables and constraints. The current leading algorithm to solve this optimization problem is NP-hard [19]. The Lagrangian relaxation of this problem can be solved more efficiently; however, the obtained solution is suboptimal because of a nonzero duality gap [13, 19].

[11]In some jurisdictions, the predicted output of large-scale renewable generators, such as wind and solar farms connected to the transmission network, is also considered in the real-time market. Short-term predictions of renewable generation are relatively accurate and, therefore, incorporating them in the real-time market could reduce the need for operating reserves.

The following section specifies the design goals for future distribution grid control mechanisms and describes a new approach to the control of a vast number of active end-nodes.

## 1.4.1 Goals

Control mechanisms for active end-nodes should prevent line and transformer overloads, mitigate large voltage fluctuations, and avoid reverse flows towards primary distribution feeders. Additionally, an admissible control must satisfy the following design goals:

- **Be legacy compatible**: Given the tremendous investments that have been made in the infrastructure of the grid, new control mechanisms should be compatible with existing components and operation rules of the grid.
- **Increase utilization**: To assure high reliability, the power system is traditionally designed and operated with a substantial operating margin.[12] The smart grid should maintain reliability while improving the utilization of generation, transmission, and distribution assets, for example by supplying elastic loads during off-peak periods.
- **Reduce carbon footprint**: Control mechanisms should support large-scale integration of low-carbon technologies into distribution networks with minimal curtailment, thereby minimizing the overall carbon footprint of the grid.
- **Be cost efficient**: The smart grid control architecture must be cost-effective. For example, it should improve the economics of demand-side technologies, thereby increasing their adoption.
- **Be fair**: End-nodes in the smart grid may differ in their types, technologies, and service requirements. In such a heterogenous system with limited available resources, fair power allocation is of paramount importance to avoid starvation. Control mechanisms should provide some notion of fairness to the end-nodes such as *proportional fairness* [17].
- **Be scalable**: Given the number of active end-nodes that will be connected to the smart grid, the underlying control system must be scalable. This is because computing a control decision that applies to these end-nodes is a computationally intensive task.
- **Be responsive**: To ensure reliability in the face of increased variability and uncertainty in the smart grid, control mechanisms should rapidly respond to contingencies and operator requests. Moreover, control mechanisms should not result in unnecessary invocation of existing protection mechanisms, which could cause service interruption and reduce the life cycle of protection equipment.
- **Be resilient**: Grid control mechanisms are expected to fail gracefully and automatically recover from a fault condition.

---

[12]For example, the notion of n-1 reliability requires the system to reliably withstand the failure of any one of its elements.

- **Be non-disruptive**: Control mechanisms should have an imperceptible impact on end users' performance, e.g., the time taken to charge an EV.

Note that a control mechanism should balance *system-level* objectives such as scalability with *user-level* objectives such as fairness. These objectives are often competing, and therefore, cannot be satisfied at the same time. A control mechanism must necessarily make trade-off between these competing objectives.

### 1.4.2  Optimal Control in Quasi Real-Time

We now discuss the broad outlines of a control scheme that meets the criteria set out above. To begin with, it is obvious that the control of active end-nodes in the distribution grid based on day-ahead predictions cannot reliably and efficiently deal with the stochastic nature of renewable generation and EV mobility. This is because control decisions that are computed based on day-ahead forecasts are very likely to be either infeasible or suboptimal at the time of their execution because of prediction errors; infeasible control decisions can put power system reliability at risk. Maintaining a conservative operating margin to accommodate these prediction errors results in low system utilization. Thus, the growing penetration of active end-nodes motivates the need for quasi real-time control based on fast timescale measurements.

Active end-nodes in the distribution network can be controlled in near real-time using two different approaches. The first approach relies on real-time measurements of the distribution network state, instead of proactive power flow calculations. Given the availability of measurement nodes in the distribution network and a reliable broadband communication network that connects them to the end-nodes, the end-nodes can learn of changes in the grid state (such as transformer and line loadings) in real-time and adjust their power consumption or production accordingly, just as the TCP endpoints in the Internet can learn of the congestion state of the network after a small delay and back off in case of congestion without having a model of the underlying network [2, 3]. However, due to the uncoordinated actions of the end-nodes, it is possible for the system to transiently move into an overload state, resulting in physical stress to grid elements such as transformers.

The second approach relies on power flow calculations, which incorporate a model of the distribution network, to compute a feasible and optimal control. An optimization problem formulated for the distribution network is solved in near real-time, using measurements of elastic and inelastic loads as well as available renewable power. This approach also requires the knowledge of real and reactive power consumption at different buses, which can be obtained through real-time measurements of the end-nodes. Chapter 4 describes this novel approach which guarantees that control is almost always admissible unlike the first approach.

## 1.5 Chapter Summary

The century-old power grid has witnessed profound changes recently due to the confluence of the following factors: (1) advances in battery and renewable technologies and the subsequent reduction in their prices, (2) introduction of high-power elastic loads, such as PEVs, into distribution systems, (3) strategic decisions made by governments to reduce reliance on fossil fuels in favor of renewable energy sources, and (4) the availability of inexpensive sensing, communication, and control devices, which paved the way for pervasive measurement and control in distribution networks. Some of these changes may subject the grid to excessive amounts of variability and uncertainty that threaten its reliability and reduce its efficiency under existing grid control paradigms. This imminent threat can be addressed by harnessing the flexibility offered by elastic loads. In particular, control of active end-nodes in quasi real-time could enable operators to meet their efficiency and fairness requirements, accommodate a higher penetration of PV generation in existing distribution systems, and enhance service reliability by preventing network overloads, reverse flows, and voltage deviations beyond operating limits.

## References

1. Abdul-Rahman K, Shahidehpour S, Aganagic M, Mokhtari S (1996) A practical resource scheduling with OPF constraints. IEEE Trans Power Syst 11(1):254–259
2. Ardakanian O, Rosenberg C, Keshav S (2013) Distributed control of electric vehicle charging. In: ACM e-Energy, pp 101–112
3. Ardakanian O, Keshav S, Rosenberg C (2014) Real-time distributed control for smart electric vehicle chargers: from a static to a dynamic study. IEEE Trans Smart Grid 5(5):2295–2305
4. Becker T, Sidhu I, Tenderich B (2009) Electric vehicles in the United States: a new model with forecasts to 2030
5. California New Car Dealers Association (2015) California auto outlook. http://www.cncda.org/CMS/Pubs/Cal_Covering_4Q_14.pdf
6. Callaway D, Hiskens I (2011) Achieving controllability of electric loads. Proc IEEE 99(1):184–199
7. Clement-Nyns K, Haesen E, Driesen J (2010) The impact of charging plug-in hybrid electric vehicles on a residential distribution grid. IEEE Trans Power Syst 25(1):371–380
8. DeForest N, Funk J, Lorimer A, Ur B, Sidhu I, Kaminsky P, Tenderich B (2009) Impact of widespread electric vehicle adoption on the electrical utility business - threats and opportunities
9. Deutsche Bank Market Research (2014) 2014 outlook: Let the second gold rush begin. http://www.deutschebank.nl/nl/docs/Solar_-_2014_Outlook_Let_the_Second_Gold_Rush_Begin.pdf
10. Electric Vehicles Initiative and International Energy Agency (2015) Global EV Outlook. http://www.iea.org/evi/Global-EV-Outlook-2015-Update_1page.pdf
11. Energiewende A (2014) The German Energiewende and its climate paradox. Report 038/04-A-2014/EN, Agora Energiewende
12. Farivar M, Neal R, Clarke C, Low S (2012) Optimal inverter VAR control in distribution systems with high PV penetration. In: IEEE PES general meeting, pp 1–7
13. Guan X, Zhai Q, Papalexopoulos A (2003) Optimization based methods for unit commitment: Lagrangian relaxation versus general mixed integer programming. IEEE PES Gen. Meet. 2:1095–1100

14. International Energy Agency (2014) CO2 emissions from fuel combustion. http://www.iea.org/publications/freepublications/publication/CO2EmissionsFromFuelCombustionHighlights 2015.pdf
15. Katiraei F, Agüero J (2011) Solar PV integration challenges. IEEE Power Energy Mag 9(3): 62–71
16. Katiraei F, Sun C, Enayati B (2015) No inverter left behind: protection, controls, and testing for high penetrations of PV inverters on distribution systems. IEEE Power Energy Mag 13(2): 43–49
17. Kelly F, Maulloo A, Tan D (1998) Rate control for communication networks: shadow prices, proportional fairness and stability. J Oper Res Soc 49(3):237–252
18. Lew D et al (2013) Wind and solar curtailment. Report NREL/CP-5500-60245, National Renewable Energy Laboratory
19. Li T, Shahidehpour M (2005) Price-based unit commitment: a case of lagrangian relaxation versus mixed integer programming. IEEE Trans Power Syst 20(4):2015–2025
20. Lopes J, Soares F, Almeida P (2011) Integration of electric vehicles in the electric power system. Proc IEEE 99(1):168–183
21. Meier A (2006) Electric Power Systems: A Conceptual Introduction. Wiley-IEEE Press, Hoboken
22. Meier A, Culler D, McEachern A, Arghandeh R (2014) Micro-synchrophasors for distribution systems. In: Innovative smart grid technologies conference (ISGT), IEEE PES, pp 1–5
23. Mills A et al (2009) Understanding variability and uncertainty of photovoltaics for integration with the electric power system. Report LBNL-2855E, Lawrence Berkeley National Laboratory
24. Nykvist B, Nilsson M (2015) Rapidly falling costs of battery packs for electric vehicles. Nature Climate Change
25. Pieltain Fernández L, Román T, Cossent R, Domingo C, Frías P (2011) Assessment of the impact of plug-in electric vehicles on distribution networks. IEEE Trans Power Syst 26(1): 206–213
26. Shao S, Pipattanasomporn M, Rahman S (2009) Challenges of PHEV penetration to the residential distribution network. In: IEEE PES general meeting, pp 1–8
27. Su X, Masoum M, Wolfs P (2014) Optimal PV inverter reactive power control and real power curtailment to improve performance of unbalanced four-wire LV distribution networks. IEEE Trans Sustain Energy 5(3):967–977
28. Taft J, De Martini P (2012) Cisco Systems – Ultra Large-Scale Power System Control Architecture. http://www.cisco.com/web/strategy/docs/energy/control_architecture.pdf
29. Taft J, De Martini P (2013) Ultra-large scale control architecture. In: IEEE PES innovative smart grid technologies (ISGT), pp 1–6
30. Tesla Motors (2016) (retrieved) Tesla Gigafactory. http://www.teslamotors.com/gigafactory
31. The Edison Foundation IEI (2014) Utility-Scale Smart Meter Deployments. http://www.edisonfoundation.net/iei/Documents/IEI_SmartMeterUpdate_0914.pdf
32. Turitsyn K, Sulc P, Backhaus S, Chertkov M (2010) Distributed control of reactive power flow in a radial distribution circuit with high photovoltaic penetration. In: IEEE PES general meeting, pp 1–6
33. Wikipedia (2015) (retrieved) LPWAN. https://en.wikipedia.org/wiki/LPWAN
34. Wood A, Wollenberg B (2012) Power generation, operation, and control, 2nd edn. Wiley, New York

# Chapter 2
# Related Work

**Abstract** The increased adoption of active end-nodes can negatively impact the reliable and economical generation, transmission, and distribution of power. This chapter gives an overview of these potential impacts and surveys related work on direct control of elastic loads to achieve both user-level and system-level objectives. Balancing these two types of objectives is nontrivial, giving rise to the design of various control architectures and many plausible control schemes as discussed in this chapter.

## 2.1 The Impact of Active End-Nodes on the Distribution Grid

Active end-nodes are becoming ubiquitous in distribution system [10, 11]. In view of this, many studies have explored the potential impacts of large-scale integration of these technologies on the electrical grid through the intensive use of steady-state and dynamic simulations [2, 24]. Performing these impact studies for a given power system is indeed quite complex owing to uncertainties about their point of connection and their size, and also the degree of correlation that might exist between loads and local renewable generation.[1]

It must be remarked that elastic loads, except for EVs, have been connected in large numbers to distribution feeders for a long time and operators have never considered them a threat to system reliability. With the availability of low cost communications in recent years, elastic loads, such as air conditioners, and space and water heaters, have been even utilized in some jurisdictions to shave the peak demand and to provide regulation service to the grid (see for example the *peaksaver* program in Ontario [22]). But unlike these loads, demands of EV chargers can be significant, and are relatively unpredictable and highly correlated. For example, EVs can be charged at up to 80 A at 240 V with AC Level 2 charging [37, 52], an instantaneous demand of 19.2 kW,

---

[1]For example, the workplace EV charging load is strongly correlated with solar generation, while the home-level EV charging load is usually correlated with wind generation.

© The Author(s) 2016
O. Ardakanian et al., *Integration of Renewable Generation and Elastic Loads into Distribution Grids*, SpringerBriefs in Electrical and Computer Engineering,
DOI 10.1007/978-3-319-39984-3_2

which is equivalent to the average demand of about ten average homes in North America. We therefore next discuss the potential impacts of EV and PV adoption on the distribution system.

### 2.1.1   Impact of EV Adoption

Studies on the impact of EV charging on distribution, transmission, and generation systems go back to the 1980s. An early paper by Heydt in 1983 [19] anticipates that the load increase due to the future penetration of EVs would fall within generation planning limits; however, distribution circuits may be inadequate to accommodate the charging of EVs; therefore, transformer overloading and voltage deviations are expected. In this regard, load management strategies are necessary to alleviate peak loading stresses. A similar observation is made by Rahman et al. [32]. The authors anticipate that with future penetration of EVs, certain distribution branches may be subject to significant overloads, even if the entire system has sufficient capacity. This is attributed to the expected nonuniform growth of the EV charging load in a distribution network.

The potential impacts of EV integration into the distribution network, including increased energy losses, transformer and branch congestion, voltage deviations that affect power quality, and phase imbalance, have been explored extensively in the literature [8, 16, 21, 27, 30, 31]. In recent work, Fernández et al. [30] assess the impact of uncontrolled EV charging on large-scale distribution system planning in two different case studies. They show that the minimum reinforcement investment required to accommodate 62 % EV penetration can increase the total network costs by up to 19 % compared to a situation without EVs. Furthermore, energy losses increase by up to 20 and 40 % of actual values in off-peak hours for 35 and 62 % EV penetration respectively. The incremental investment can be reduced by 60–70 % if a smart charging strategy is adopted.

In an effort to underscore the need for coordinated charging, Qian et al. [31] analyze the impact of four different EV charging strategies on a typical UK distribution system. In the case of uncontrolled domestic charging, where EVs start charging nearly simultaneously, the daily peak load increases by 17.9 and 35.8 % for 10 and 20 % EV market penetration levels respectively. This drastic increase in the peak load overloads several branches and transformers, emphasizing the need for control.

In a similar line of work, Lopes et al. [27] evaluate the impact of EV integration into a typical medium voltage distribution network in terms of branch congestion levels and voltage profiles for different charging strategies. The authors show that the voltage lower limit is almost reached at several distant buses in the scenario with 10 % EV penetration and uncoordinated charging. However, the lower voltage limit is reached only when EV penetration reaches 52 % if a smart charging strategy is adopted. The branch congestion level, i.e., the ratio of the line loading to its rating, is only slightly higher for the case of 52 % EV penetration and smart charging than the case of 10 % EV penetration and uncoordinated charging, indicating the effectiveness of a smart charging strategy in relieving congestion.

Clement Nyns et al. [8] also study the impact of low to moderate EV penetration on distribution system losses and voltage deviations. Their results imply that with 30 % EV penetration, uncoordinated charging leads to more than 10 % voltage deviations, whereas coordinated charging keeps voltage deviations below 10 % at all times. Moreover, for all charging periods and seasons, power losses noticeably decrease with coordinated charging.

The impact of EV adoption on aging of distribution transformers is explored by Gong et al. [16] and Hilshey et al. [21]. In [16], a transformer thermal model is used to study the impact of Level 2 EV charging on aging of the distribution transformers installed at residential neighbourhoods. Monte Carlo simulation results show that with poor coordination of charging times, the transformer insulation life is greatly affected at relatively high EV penetration rates. Simulation results presented in [21] indicate that coordinated charging of EVs can reduce the annual transformer aging rate by more than 12.8 and 48.9 % compared to uncoordinated charging when EV chargers are Level 1 and Level 2 (as established in [37]), respectively.

The above studies show that uncoordinated charging of a large population of EVs could have detrimental impacts on the existing distribution networks. Upgrading distribution circuits alone, would be quite costly for DSOs as discussed in [30]. Therefore, DSOs must incorporate a control strategy to reduce the required distribution reinforcement investment. Some of these control strategies are discussed in Sect. 2.2.

### 2.1.2  Impact of PV Adoption

The exponential growth of global PV cumulative installed capacity [11] has given impetus to the study of solar integration into power distribution networks and of the resultant architectural, technical, and operational problems, such as adverse impacts on power quality, protection coordination, voltage profiles, and feeder and transformer loading [24, 49]. The potential steady-state and transient impacts of PV systems on volt/var control, power quality, and power system operation depend on the penetration level and interconnection of PV units, and their interactions with loads and distribution equipment, making it extremely complex to evaluate these impacts.

Several attempts have been made to quantify the extent of local and system-wide problems associated with PV integration. In [45], it has been shown using simulations of a test network with rooftop PV systems connected to secondary distribution lines that a 30 % penetration of PV systems can be accommodated without any change to voltage control systems. Should the PV penetration increase to 50 %, over-voltage is observed in simulations; this suggests that the voltage control systems must be adjusted or re-engineered at this penetration. Another study examines the impacts of high penetration of residential PV systems on distribution system protection and voltage control [5]. The conclusion is that high PV penetration complicates the coordination of protection equipment and creates unacceptable voltage swings (beyond pre-defined limits) on feeders. In [20], a control methodology is described for grid-scale battery storage systems to address the negative impacts of PV integration; this methodology enables storage systems to provide voltage stability and frequency regulation, and improves the economics of distributed solar generation.

## 2.2 Control of EV Chargers

The control of smart EV chargers is essential to address the potential distribution network problems discussed in Sect. 2.1, while satisfying user-level objectives. Additionally, smart EV chargers and other elastic loads can be controlled to support higher penetrations of distributed renewable generation, achieve a desired response to power system dynamics, or provide system services such as frequency regulation [1, 6, 15, 17, 25, 42]. This section only surveys control mechanisms that aim to mitigate the negative impacts of EVs on the distribution network and to optimize certain user-level objectives; thus, it does not touch upon control mechanisms for delivering electricity to the grid in vehicle-to-grid (V2G) applications.

The main focus of this section is on control schemes that do not put customers in the control loop, meaning that the customers may specify charging deadlines and preferences but cannot impede or delay the execution of control decisions that are computed by the utility based on their input. These schemes are referred to as *direct control* schemes. Unlike direct load control schemes, price-based schemes assume some specific response from the customers to changes in the electricity price. This assumption does not necessarily hold in practice and the demand response is neither predictable nor immediate, rendering these schemes of limited practical value [6]. For this reason, these control schemes are not discussed in this section.

Existing work can be categorized into two based on the objectives they seek to achieve. The first category encompasses approaches that take the perspective of the electric utility and satisfy one or several system-level objectives, whereas the second category encompasses approaches that take the perspective of users and satisfy one or several user-level objectives, where users are either EV owners or charge service providers (CSPs). These control objectives can be myopic or defined over a finite or infinite time horizon. Furthermore, there are several possible approaches to control EV chargers. In particular, EV chargers have been controlled using a schedule computed the prior day (known as *pre-dispatch scheduling*) or in near real-time. Control decisions can be made independently by EV chargers (a fully distributed approach), jointly by EV chargers and intermediate control nodes installed at transformers (a decentralized approach), or entirely by a computer cluster at the utility control center (a centralized approach). Finally, the control scheme may require the precise model of the distribution network along with load and generation forecasts, or only rely on recent measurements of certain network parameters. Thus, the extensive body of literature that has been developed around the control of elastic loads can be divided into several categories based on the following criteria:

**Table 2.1** Taxonomy of related work

| | Objective | Pre-dispatch | Real-time | |
|---|---|---|---|---|
| | | | Centralized | Decentralized |
| Utility | Avoid network congestion | [36] | [34, 35] | [18, 21, 50] |
| | Improve voltage profiles | [8, 48] | [4, 9] | |
| | Minimize losses | [8, 36, 38] | [9] | |
| | Flatten the load | [14, 38] | | [14, 26] |
| | Shave the peak load | [29, 43] | [46] | [12] |
| | Minimize the cost of generation | | [9] | |
| Users | Minimize charging cost | [28, 33, 36, 43] | [51] | [23] |
| | Minimize charging time | | [53] | [39] |
| | Maximize EV owners' convenience | | [50] | [50] |
| | Maximize CSP's revenue | | [7] | |
| | Fair power allocation to EVs | | [40] | [12] |

- **Time of control**: The control algorithm can run in near real-time or several hours in advance of power delivery.
- **Information needs**: The control algorithm may require the precise model of the distribution network along with load and generation forecasts, or rely on recent measurements of certain network parameters only.
- **Decision-making approach**: Control decisions can be computed in a centralized or decentralized manner. The computation and communication overhead of control greatly depends on this.
- **Optimization horizon**: Control objectives can be myopic or defined over a finite or infinite time horizon.

Table 2.1 shows a taxonomy of existing work on coordinated charging according to their objectives and control approaches.

## 2.2.1 Pre-Dispatch Scheduling

Pre-dispatch scheduling approaches compute charging schedules for EVs by solving an optimization problem in advance of power delivery. In some cases, this optimization problem falls within the general class of optimal power flow problems [8, 29, 36, 48]. Solving the OPF problem requires a precise model of the distribution network and inelastic loads, as well as the knowledge of the point of connection of chargers and arrival and departure patterns of EVs. These parameters are difficult to determine or estimate in practice. Hence, pre-dispatch scheduling approaches either maintain a conservative operating margin to accommodate estimation uncertainties or perform power flow calculations for numerous instantiations of random variables, e.g., EV arrival and departure times, their initial state of charge (SOC), and their point of connection. The former typically results in system under-utilization and the latter significantly increases the computation time.

For example, Mehboob et al. [29] solve a distribution optimal power flow (DOPF) problem to determine the hourly EV charging schedule and hourly tap and capacitor settings that minimize the system peak. This optimization problem incorporates voltage and feeder capacity constraints as well as EV charging constraints. They employ a genetic algorithm based approach to solve this DOPF problem. This approach generates many feasible EV load samples and performs power flow calculations for each set of samples to find a day-ahead most likely solution.

A DOPF problem is also formulated in [36] to control EV charging loads, taps, and capacitor switching decisions for the next day in an unbalanced three-phase distribution system. The authors consider various objectives and incorporate the distribution substation capacity constraint in the nonlinear programming problem. Specifically, they minimize the total energy drawn by the local distribution company and its cost, the total feeder losses, and the total cost of EV charging over the period of a day. The proposed day-ahead hourly scheduling approach is evaluated on the IEEE 13-node test feeder and a real distribution feeder. Compared to the uncontrolled charging case, their approach prevents undervoltage conditions and reduces the peak demand and losses.

Several other pre-dispatch scheduling approaches simply use the optimal control framework without relying on power flow calculations. In these cases, control might be computed more efficiently; however, it does not necessarily respect distribution network constraints such as voltage limits. Following is an overview of the most relevant work in this area.

In recent work, Gan et al. [14] and Ma et al. [28] use distributed control to obtain a day-ahead charging schedule for EVs. In [14], the EV charging control problem is formulated as a discrete optimization problem with the objective of flattening the aggregate demand served by a transformer. A stochastic decentralized control algorithm is proposed to find an approximate solution to this optimization problem. It is shown that this algorithm almost surely converges to one of the equilibrium charging profiles. To facilitate real-time implementation of this controlled charging scheme, the authors also propose an online version of their decentralized control algorithm in which EVs participate in negotiation on their charging profiles as they plug in for charging, over time. In [28], a decentralized algorithm is proposed to find the EV charging strategy that minimizes individual charging costs. It is shown that the optimal strategy obtained using this algorithm converges to the unique Nash equilibrium strategy when there is an infinite population of EVs. In the case of homogeneous EV populations, this Nash equilibrium strategy coincides with the valley-filling maximizing strategy, i.e., the globally optimal strategy.

In [43], a deterministic optimization problem is formulated to find a fleet charging schedule which minimizes the overall charging cost, subject to the available power, the battery capacity, and the charging power constraints. The optimization problem is deterministic because it is assumed that the connection and disconnection times of EVs, their energy demands, the price of electricity, and the total wind generation are known a priori. The authors also compare linear and quadratic approximations of the EV battery behavior in terms of violations of the battery boundaries (minimum and maximum charge levels) for the obtained charging schedule.

A similar line of work by Rotering et al. [33] explores the possibility of using plug-in hybrid EVs for regulation and ancillary services while charging their batteries with minimum cost. Specifically, dynamic programming is employed to find a charging schedule that minimizes the EV charging cost based on forecasts of the electricity price, EV driving patterns, and energy demands in three different scenarios. If the control is incapable of supplying the energy demand, it is assumed that the lack of charge is fulfilled by the internal combustion engine consuming gas, which is presumably more expensive than electricity.

The relationship between the objectives that are based on load factor, load variance, and losses is investigated in [38]. The authors formulate three optimization problems to minimize the load variance, to maximize the load factor, and finally to minimize losses. These optimization problems are solved by a centralized approach using day-ahead load predictions, noting that the first two problems are convex and can be solved more efficiently compared to the third one. It is shown through simulations on practical distribution systems that solutions to these three problems are close, motivating the use of load factor or load variance as the objective function rather than losses. In any case, for practical systems the performance of the algorithm that minimizes the load variance is quite similar to the one that minimizes losses.

### 2.2.2 Near Real-Time Control

The near real-time computation of charging schedules improves utilization and reliability of the power system compared to the pre-dispatch computation of the schedules by continuously adapting the charging rate of EV chargers to the available capacity of the network. Hence, smart EV chargers use higher rates when the distribution network has sufficient capacity, reducing these rates once the network becomes congested. The real-time charging schedule could be computed using either a centralized or a decentralized/distributed approach. The following surveys related work in each category.

#### 2.2.2.1 Centralized Approaches

Coordinated charging of EVs at parking facilities with a maximum total available amount of power is the focus of [7, 40, 51, 53]. In [53], the problem of finding a schedule in a charging station with stochastic EV arrivals, variable electricity prices, and intermittent renewable generation is modelled as a constrained stochastic optimization problem which can be studied using the Markov decision process framework. The objective is to minimize the mean waiting time of EVs. In [51], the scheduling problem of EV charging with stochastic arrivals and renewable generation is formulated as an infinite-horizon Markov decision process. The objective is to maximize a social welfare function that takes into account the total utility of customers, the electricity cost associated with the charging schedule, and the penalty for failing to meet the deadlines. In [7], it is assumed that there is a CSP that uses collocated renewable sources and supplements the renewable with the energy purchased from the grid. The authors formulate an online scheduling problem with the objective of maximizing the operating profit of the CSP while meeting the charging deadlines. This optimization problem is a mixed integer program. A subset of EVs are selected for charging through an admission control process and admission decisions are made based on EV arrivals, output of renewable sources, and the electricity price. The scheduler can further optimize on the time and quantity of the energy purchased from the grid.

In an effort to provide a notion of fairness, an optimization problem is formulated in [40] that maximizes a weighted average of the state of charge of parked EVs in the next time step, subject to the amount of energy available from the utility, the maximum energy that can be absorbed by EVs, and the ramp rate of EV batteries. Each weight term incorporated in the objective function is a function of the energy price, and the remaining charging time and the present SOC of the corresponding EV. The authors use four computational intelligence-based algorithms, namely the estimation of distribution algorithm, the particle swarm algorithm, the genetic algorithm, and the interior point method, to solve this optimization problem and compare their performance.

A DR strategy is proposed in [34, 35] to avoid transformer and feeder overloads by controlling non-critical and controllable loads, including EVs. This strategy determines household demand limits using a simple algorithm which protects the distribution network from congestion, and issues the obtained limits to in-home controllers. Subsequently, every in-home controller determines which appliances should be on based on the priorities and preferences set by users in advance. The effect of the proposed DR on consumers comfort is quantified using comfort indices introduced in [35]. Nevertheless, the proposed DR strategy does not guarantee congestion prevention because appliances might be turned on to satisfy users' preferences even when the transformer is congested.

Deilami et al. [9] propose a real-time smart load management algorithm to coordinate EV chargers; this algorithm minimizes the total cost of generation and anticipated losses, while respecting user preferences. To solve this problem, their approach is to use the maximum sensitivities selection method, which selects EVs for charging from a queue sorted based on the sensitivity of the loss function to EV charging loads. A load flow analysis is performed in each time step to evaluate the objective function and ensure that system constraints, including voltage limits and the available generation capacity, are not violated. Nevertheless, this approach does not deal with the distribution network problems, such as line and transformer congestion, and does not provide fairness.

A two-stage controller based on a model predictive control (MPC) formulation is designed in [4] to regulate charging of a time-varying number of EVs and control a fixed number of distributed generation inverters under the assumption that the load is periodic with period length of 24 h. Using approximate power flow equations for radial distribution networks, the proposed controller charges EV batteries to a desired SOC while tracking an optimal reachable reference voltage at every bus. The proposed scheme handles plug-and-play charging requests (as EVs join or leave the system) by updating reference voltages to ensure stability and reliability under the new dynamics. This plug-and-play operation comes at the price of delaying charging of EVs that have arrived recently until bus voltages converge to the updated reference values. Note that this control scheme does not address branch and transformer congestion problems in the distribution network.

Turitsyn et al. [46] aim at maximizing the utilization of the excess distribution circuit capacity while keeping the probability of a circuit overload negligible by controlling EV chargers. Using one-way broadcast communication, the authors regulate EV charging start times by computing a single EV connection rate and sending it periodically to the chargers. This rate determines, on average, how many EV chargers can start charging per unit time.

In summary, most existing work on real-time centralized control of EV chargers suffers from a scalability problem since computing the charging schedule for a vast number of connected EVs is computationally expensive in a centralized fashion. Moreover, centralized control schemes may require communication of sensitive information, such as EV departure times, to a central controller. The central controller is also a single point of failure in the distribution network. These issues can be addressed by distributing control among the EV chargers and possibly other control nodes as suggested in [44]. Real-time decentralized control schemes are reviewed next.

### 2.2.2.2 Decentralized Approaches

This section describes decentralized and fully distributed control schemes that run in near real-time. These schemes are scalable, robust, and use real-time information instead of long-term predictions. However, they suffers from three major shortcomings. First, they do not use a realistic model of the distribution network, which includes all branches and transformers and their operational constraints. Second, they do not evaluate the proposed solution using power flow analysis when it is not originally found using power flow calculations. Instead, many of them focus on flattening the demand of the entire distribution network, ignoring bus voltages and loading of distribution lines and transformers. Third, they do not balance efficiency and fairness of the control algorithm. In fact, fairness is not a design goal of most of these approaches. These schemes are discussed in the following.

Wen et al. [50] propose a novel approach to the EV charging control problem, where a subset of EVs are selected for charging in every time slot such that user convenience is maximized and branch flow constraints are met. This selection problem is posed as a combinatorial optimization problem, whose convex relaxation can be solved in a control center using linear programming. An efficient decentralized algorithm is then proposed based on the alternating direction method of multipliers to determine the set of EVs that must be charged in a given time slot. Using numerical simulations for different EV penetration levels, the proposed centralized and decentralized approaches are compared in terms of performance, computational complexity, and communication overhead. The authors study the effects of the control timescale and the rounding method, which maps continuous selection variables into 0 and 1, on the performance of the decentralized algorithm. Nevertheless, this paper only addresses branch congestion and ignores other operational constraints of the distribution network, does not use power flow analysis to validate that computed charging schedules are feasible, and finally does not attempt to allocate power to connected EVs based on a well-established notion of fairness.

Fan [12] borrows the notion of congestion pricing from the Internet to reduce the peak load while providing weighted proportional fairness to end users. Exploiting two-way communications between the utility and users, congestion prices are sent to users, enabling them to adapt their demands to the capacity of the market in a fully distributed fashion. The user preference is modelled as a willingness-to-pay parameter, i.e., the weight factor in the utility function of users. The proposed algorithm is then applied to EV charging to obtain a charging rate allocation. Interestingly, the total EV charging load varies with the range from which the weight factors can be chosen. Thus, the utility has to limit this range to ensure that the total load is not greater than the market capacity. Convergence behavior of the algorithm is studied using both an analytical approach and a simulation-based approach. Note that this work does not model the distribution network and does not incorporate the capacity constraints of distribution lines and transformers and the charge rate constraints of EV chargers.

In [39], several additive-increase multiplicative-decrease (AIMD) based algorithms are used for distributed control of a set of EV chargers that share a single constrained resource. The EV chargers independently increase their demands by an additive factor until the shared resource becomes congested; following this event, they reduce their demands by a multiplicative factor to relieve congestion. The authors study the problem from the user perspective rather than the utility perspective; they consider various scenarios and user-level objectives, and propose an AIMD-like congestion control algorithm for each scenario. Moreover, this work does not investigate the potential distribution network problems and is not based on the theory of network utility maximization (NUM); it instead relies on an arbitrary choice of AIMD parameters.

In [18] a control mechanism is designed to deal with transformer overloading by modelling the transformer thermal limit as a constraint. Specifically, the authors formulate the EV charging problem as an open-loop centralized control problem with the objective of minimizing the SOC deviations from 100 % and also minimizing the control effort subject to the capacity constraint of batteries and EV chargers, the temperature constraint of the substation transformer, and the target SOC specified by EV owners. Using the dual decomposition method, an iterative price-coordinated implementation of this control mechanism is proposed which allows EV owners to compute their charging rate locally. A receding-horizon feedback mechanism is employed to account for unexpected disturbances, including fluctuations in inelastic demands, changes in the number of connected EVs, changes in the ambient temperature, and modelling errors. Note that this work deals with the substation transformer overloading problem and cannot prevent distribution line overloads. Furthermore, it does not use power flow analysis to validate the operation of the proposed algorithm in a test distribution network.

Li et al. [26] aim at flattening the load at the distribution network level by extending the "max-weight algorithm" to the EV charging control problem. Control rules solve an optimization problem that minimizes the L2 norm of the aggregate load. It is shown that, in the long term, the solution to this optimization problem can be made arbitrarily close to the solution of the optimization problem that minimizes the variance of the aggregate load. The former problem can be solved in real-time. Using numerical simulations of the IEEE 37-bus and 123-bus test feeders, the performance of the algorithm is compared with static charging algorithms that use perfect knowledge and imperfect forecast of the base load for different penetration levels. Note that the authors do not attempt to address the distribution network problems due to the simultaneous charging of EVs and their objective is merely to flatten the load.

Jin et al. [23] propose an EV charging scheduling algorithm to minimize the energy bill of users, and, at the same time, flatten the aggregated load imposed on the power grid. The authors employ a grouping algorithm and a sliding window iterative scheduling algorithm. The grouping algorithm reduces the computation and communication overhead of the scheduling algorithm. It runs at a centralized coordinator, which is called the information center, and classifies the EV population into several groups based on a similarity metric defined in terms of the start and end charging times, the energy requirement of an EV, and the maximum charge rate of its charger. Once EV groups are formed, the information center computes and broadcasts the charging characteristic of every group. The charging schedule is then computed using a sliding window iterative scheduling algorithm. Specifically, EVs belonging to each group solve an optimization problem to minimize the group bill and compute their charging schedule locally in a specific slot of every cycle, while charging schedules of other groups remain unchanged. When the charging rates of EVs within a group are determined, they send their updated charging profiles to the information center. The information center broadcasts real-time price/load information at the beginning of each slot of each cycle to coordinate EV chargers. When the algorithm converges, the obtained charging schedule also optimizes the total generation cost. This work does not take fairness into account, does not address distribution network problems due to the simultaneous charging of EVs, and does not validate the results through power flow analysis of a test distribution system.

Hilshey et al. [21] propose two automaton-based strategies for coordinating EV charging to limit the power supplied by transformers and decelerate their aging.

Their approach is to compare the transformer aging status against four thresholds to determine whether the number of EVs being charged should be increased, decreased, or held constant in the next time period. Once the number of active chargers is determined, one of the proposed decentralized automaton-based strategies is used for admission control. The first strategy is a first-come first-served strategy in which every EV sends a charge request to the transformer. If the request is denied due to congestion, it is queued and processed again in the next time slot. The second strategy is probabilistic; it allows chargers to specify urgency by agreeing to pay a higher rate. If charging is not urgent and the request is denied, the charge request is sent to the transformer in the next time slot with a probability $p$. If charging is urgent and the request is denied, a request is sent to the transformer in every future time slot until it is admitted. Note that both control strategies do not provide fairness, and are only applicable to a single transformer supplying fixed-rate EV chargers. Moreover, the aging thresholds are chosen in an ad hoc manner.

## 2.3   Control of Renewable Inverters

Control will not be limited to elastic loads in the smart grid. Smart inverters, which are capable of injecting and consuming reactive power and curtailment of real power, can also be controlled by the utility to address growing concerns over widespread adoption of PV systems, and also achieve several system-level objectives. The optimal control of PV inverters has received increased attention in recent years. For example, Farivar et al. [13] propose the fast timescale control of the reactive power injection of PV inverters to minimize line and inverter losses as well as the energy consumption through voltage optimization. This problem is formulated as an OPF for a radial distribution system and the optimal voltage regulation operation is evaluated on a distribution feeder on the Southern California Edison system. Similarly, the authors of [3] propose a real-time distributed control of the reactive power output of smart inverters to minimize feeder head real power consumption. They utilize a model-free control algorithm that relies on periodic measurements of the feeder head real power broadcast by the substation.

An OPF problem is formulated in [41] to determine an PV inverter control strategy that improves voltage magnitude and balance profiles, while minimizing network losses, inverter losses, and solar generation curtailment. This multi-objective optimization problem is solved using a sequential quadratic programming approach. The performance of this control strategy is evaluated through power flow analysis in a real unbalanced three-phase low-voltage distribution system in Australia. In a similar line of work, Turitsyn et al. [47] find the optimal dispatch of the inverter's reactive power to minimize line losses and maintain the voltage within an acceptable range in a radial distribution system. None of these control schemes takes advantage of elastic loads to minimize the curtailment of solar power.

## 2.4 Joint Control of Elastic Loads and Renewable Energy Systems

The impact studies presented in Sect. 2.1 suggest that PV systems and elastic loads might have opposite impacts on distribution circuits. Specifically, the uncontrolled charging of a large population of EVs can result in undervoltage and equipment overloading, whereas the uncontrolled operation of a large number of solar inverters can cause overvoltage and reverse power flow toward the distribution substation. Thus, it is reasonable to extend the optimal control framework to jointly control smart inverters and elastic loads. This control scheme enables the grid to safely accommodate higher penetrations of renewable generation and elastic loads, while enhancing the overall reliability and cost-effectiveness of the power system. Despite the significance of such a control scheme in future distribution systems, the synergy between inverter-based renewable energy systems and elastic loads has not been exploited in the literature to stabilize voltage, relieve congestion, prevent reverse flows, and minimize curtailment in a distribution system with a high concentration of renewable generation. Chapter 4 expands on this idea.

## 2.5 Chapter Summary

Large-scale integration of elastic loads and renewable energy systems can negatively impact reliable and economical generation, transmission, and distribution of power if these end-nodes are not controlled properly. This has given impetus to the design of mechanisms to control elastic loads and renewable inverters. However, most related work focuses on controlling elastic loads, very little work focuses on controlling smart inverters, and practically no work explores the joint control of these technologies. The extensive body of literature that has been developed around the control of elastic loads can be divided into several categories based on the following criteria:

- **Time of control:** The control algorithm can run in near real-time or several hours in advance of power delivery.
- **Information needs:** The control algorithm may require the precise model of the distribution network along with load and generation forecasts, or rely on recent measurements of certain network parameters only.
- **Decision-making approach:** Control decisions can be computed in a centralized or decentralized manner. The computation and communication overhead of control greatly depends on this.
- **Optimization horizon:** Control objectives can be myopic or defined over a finite or infinite time horizon.

A control scheme that fully meets the design goals specified in Chap. 1 must be decentralized and based on real-time measurements. This scheme should enhance power system reliability and efficiency, reduce its carbon emissions, satisfy user-level objectives, and mitigate adverse impacts of large-scale adoption of solar PV systems and EVs, including large voltage fluctuations, network congestion, reverse flow, and violation of voltage limits. None of the control schemes surveyed in this chapter can satisfy all of these objectives while meeting the design goals specified in the previous chapter.

# References

1. Ahn C, Li C, Peng H (2011) Optimal decentralized charging control algorithm for electrified vehicles connected to smart grid. J Power Sources 196(2):10,369–10,379
2. Amrhein M, Krein P (2005) Dynamic simulation for analysis of hybrid electric vehicle system and subsystem interactions, including power electronics. IEEE Trans. Veh. Technol. 54(3): 825–836
3. Arnold DB, Negrete-Pincetic M, Stewart EM, Auslander DM, Callaway DS (2015) Extremum Seeking control of smart inverters for VAR compensation. In: IEEE Proceedings of the PES general meeting, pp 1–5
4. Bansal S, Zeilinger M, Tomlin C (2014) Plug-and-play model predictive control for electric vehicle charging and voltage control in smart grids. In: Proceedings of the IEEE conference on decision and control, pp 5894–5900
5. Baran M, Hooshyar H, Shen Z, Huang A (2012) Accommodating high PV penetration on distribution feeders. IEEE Trans. Smart Grid 3(2):1039–1046
6. Callaway D, Hiskens I (2011) Achieving controllability of electric loads. Proc. IEEE 99(1): 184–199
7. Chen S, Tong L (2012) IEMS for large scale charging of electric vehicles: architecture and optimal online scheduling. In: IEEE Proceedings of the smart grid communications, pp 629–634
8. Clement-Nyns K, Haesen E, Driesen J (2010) The impact of charging plug-in hybrid electric vehicles on a residential distribution grid. IEEE Trans. Power Syst. 25(1):371–380
9. Deilami S, Masoum A, Moses P, Masoum MAS (2011) Real-time coordination of plug-in electric vehicle charging in smart grids to minimize power losses and improve voltage profile. IEEE Trans. Smart Grid 2(3):456–467
10. Electric Vehicles Initiative and International Energy Agency Global EV Outlook (2015). http://www.iea.org/evi/Global-EV-Outlook-2015-Update_1page.pdf
11. EPIA (2014) Global market outlook for photovoltaics 2014–2018. http://www.epia.org/news/publications/global-market-outlook-for-photovoltaics-2014-2018/
12. Fan Z (2012) A distributed demand response algorithm and its application to PHEV charging in smart grids. IEEE Trans. Smart Grid 3(3):1280–1290
13. Farivar M, Neal R, Clarke C, Low S (2012) Optimal inverter VAR control in distribution systems with high PV penetration. In: Proceedings of the IEEE PES general meeting, pp 1–7
14. Gan L, Topcu U, Low S (2013) Optimal decentralized protocol for electric vehicle charging. IEEE Trans. Power Syst. 28(2):940–951
15. Gan L, Wierman A, Topcu U, Chen N, Low S (2013) Real-time deferrable load control: handling the uncertainties of renewable generation. In: Proceedings of the ACM e-Energy, pp 113–124
16. Gong Q, Midlam-Mohler S, Marano V, Rizzoni G (2012) Study of PEV charging on residential distribution transformer life. IEEE Trans. Smart Grid 3(1):404–412
17. Hao H, Sanandaji B, Poolla K, Vincent T (2015) Aggregate flexibility of thermostatically controlled loads. IEEE Trans. Power Syst. 30(1):189–198
18. Hermans R, Almassalkhi M, Hiskens I (2012) Incentive-based coordinated charging control of plug-in electric vehicles at the distribution-transformer level. In: Proceedings of the American control conference (ACC), pp 264–269

19. Heydt G (1983) The impact of electric vehicle deployment on load management straregies. IEEE Trans. Power Appar. Syst. PAS-102(5):1253–1259
20. Hill C, Such M, Chen D, Gonzalez J, Grady W (2012) Battery energy storage for enabling integration of distributed solar power generation. IEEE Trans. Smart Grid 3(2):850–857
21. Hilshey A, Rezaei P, Hines P, Frolik J (2012) Electric vehicle charging: Transformer impacts and smart, decentralized solutions. In: Proceedings of the IEEE PES general meeting, pp 1–8
22. HydroOne (2015, retrieved) peaksaver PLUS. http://www.hydroone.com/peaksaver
23. Jin R, Wang B, Zhang P, Luh P (2013) Decentralised online charging scheduling for large populations of electric vehicles: a cyber-physical system approach. Int. J. Parallel, Emerg. Distrib. Syst. 28(1):29–45
24. Katiraei F, Agüero J (2011) Solar PV integration challenges. IEEE Power Energy Mag. 9(3): 62–71
25. Kempton W, Tomic J (2005) Vehicle-to-grid power implementation: from stabilizing the grid to supporting large-scale renewable energy. J Power Sources 144(1):280–294
26. Li Q, Cui T, Negi R, Franchetti F, Ilic M (2011) On-line decentralized charging of plug-in electric vehicles in power systems
27. Lopes J, Soares F, Almeida P (2011) Integration of electric vehicles in the electric power system. Proc. IEEE 99(1):168–183
28. Ma Z, Callaway D, Hiskens I (2013) Decentralized charging control of large populations of plug-in electric vehicles. IEEE Trans. Control Syst. Technol. 21(1):67–78
29. Mehboob N, Canizares C, Rosenberg C (2014) Day-ahead dispatch of PEV loads in a residential distribution system. In: Proceedings of the IEEE PES general meeting, pp 1–5
30. Pieltain Fernández L, Román T, Cossent R, Domingo C, Frías P (2011) Assessment of the impact of plug-in electric vehicles on distribution networks. IEEE Trans. Power Syst. 26(1):206–213
31. Qian K, Zhou C, Allan M, Yuan Y (2011) Modeling of load demand due to EV battery charging in distribution systems. IEEE Trans. Power Syst. 26(2):802–810
32. Rahman S, Shrestha G (1993) An investigation into the impact of electric vehicle load on the electric utility distribution system. IEEE Trans. Power Deliv. 8(2):591–597
33. Rotering N, Ilic M (2011) Optimal charge control of plug-in hybrid electric vehicles in deregulated electricity markets. IEEE Trans. Power Syst. 26(3):1021–1029
34. Shao S, Pipattanasomporn M, Rahman S (2011) Demand response as a load shaping tool in an intelligent grid with electric vehicles. IEEE Trans. Smart Grid 2(4):624–631
35. Shao S, Pipattanasomporn M, Rahman S (2012) Grid integration of electric vehicles and demand response with customer choice. IEEE Trans. Smart Grid 3(1):543–550
36. Sharma I, Canizares C, Bhattacharya K (2014) Smart charging of PEVs penetrating into residential distribution systems. IEEE Trans. Smart Grid 5(3):1196–1209
37. Society of Automotive Engineers (2016, retrieved) SAE J1772 Standard. http://www.sae.org/smartgrid/chargingspeeds.pdf
38. Sortomme E, Hindi M, MacPherson S, Venkata S (2011) Coordinated charging of plug-in hybrid electric vehicles to minimize distribution system losses. IEEE Trans. Smart Grid 2(1):198–205
39. Studli S, Crisostomi E, Middleton R, Shorten R (2012) AIMD-like algorithms for charging electric and plug-in hybrid vehicles. In: Proceedings of the IEEE international electric vehicle conference (IEVC), pp 1–8
40. Su W, Chow M (2012) Computational intelligence-based energy management for a large-scale PHEV/PEV enabled municipal parking deck. Appl. Energy 96:171–182
41. Su X, Masoum M, Wolfs P (2014) Optimal PV inverter reactive power control and real power curtailment to improve performance of unbalanced four-wire LV distribution networks. IEEE Trans. Sustain. Energy 5(3):967–977
42. Subramanian A, Garcia M, Dominguez-Garcia A, Callaway D, Poolla K, Varaiya P (2012) Real-time scheduling of deferrable electric loads. In: Proceedings of the American control conference (ACC), pp 3643–3650
43. Sundström O, Binding C (2010) Optimization methods to plan the charging of electric vehicle fleets. In: Proceedings of the international conference on control, communication and power engineering, pp 28–29

44. Taft J, De Martini P (2012) Cisco Systems–Ultra Large-Scale Power System Control Architecture. http://www.cisco.com/web/strategy/docs/energy/control_architecture.pdf
45. Thomson M, Infield D (2007) Network power-flow analysis for a high penetration of distributed generation. IEEE Trans. Power Syst. 22(3):1157–1162
46. Turitsyn K, Sinitsyn N, Backhaus S, Chertkov M (2010) Robust broadcast-communication control of electric vehicle charging. In: Proceedings of the IEEE smart grid communications, pp 203–207
47. Turitsyn K, Sulc P, Backhaus S, Chertkov M (2010) Distributed control of reactive power flow in a radial distribution circuit with high photovoltaic penetration. In: Proceedings of the IEEE PES general meeting, pp 1–6
48. Vlachogiannis J (2009) Probabilistic constrained load flow considering integration of wind power generation and electric vehicles. IEEE Trans. Power Syst. 24(4):1808–1817
49. Walling R, Saint R, Dugan R, Burke J, Kojovic L (2008) Summary of distributed resources impact on power delivery systems. IEEE Trans. Power Deliv. 23(3):1636–1644
50. Wen C, Chen J, Teng J, Ting P (2012) Decentralized plug-in electric vehicle charging selection algorithm in power systems. IEEE Trans. Smart Grid 3(4):1779–1789
51. Xu Y, Pan F (2012) Scheduling for charging plug-in hybrid electric vehicles. In: Proceedings of the IEEE conference on decision and control, pp 2495–2501
52. Yilmaz M, Krein P (2013) Review of battery charger topologies, charging power levels, and infrastructure for plug-in electric and hybrid vehicles. IEEE Trans. Power Electron. 28(5): 2151–2169
53. Zhang T, Chen W, Han Z, Cao Z (2014) Charging scheduling of electric vehicles with local renewable energy under uncertain electric vehicle arrival and grid power price. IEEE Trans. Veh. Technol. 63(6):2600–2612

# Chapter 3
# System Model

**Abstract** This chapter presents models and operating constraints for inelastic loads, PV systems, EV chargers, and dedicated storage systems connected to distribution feeders, along with a linear branch flow model for power flow analysis in radial distribution systems. These models can be used to formulate a control problem for active end-nodes. A plausible fairness criterion is also introduced in this chapter.

## 3.1 Power Distribution System

The power distribution system comprises a large number of lines, transformers, and other devices that are essential for reliable delivery of electricity to customers in urban and rural areas [12]. A radial distribution system[1] typically has a single source of supply, i.e., the distribution substation, delivering power to residential and commercial loads through feeders radiating from the substation and *laterals* (or secondary distribution lines) branching from these feeders at certain points, known as *buses*. Figure 3.1 depicts the one-line diagram[2] of a three-phase distribution network interfacing with the transmission network and power stations at the substation. The voltage is initially reduced by the substation transformer and later by pad-mount and pole-top transformers, which feed a small number of customers in a neighbourhood, to the nominal supply voltage.

A radial distribution system has a logical tree topology. The substation is the *root* of this tree, and electrical loads, such as homes and businesses, are its *leaves*. Radial systems have been designed with the assumption that real power flows always in the same direction, from the substation to loads. Reverse power flow can negatively affect the operation of voltage regulators and protective devices [15], and is therefore not allowed in the distribution system beyond *balancing zone*.

---

[1]Most distribution systems are radial. Even in some cases where the network topology is a mesh, switches are often operated in a way that power flows only on a radial sub-graph of the network.

[2]A one-line diagram represents all phase conductors between two buses by a single line.

© The Author(s) 2016

O. Ardakanian et al., *Integration of Renewable Generation and Elastic Loads into Distribution Grids*, SpringerBriefs in Electrical and Computer Engineering, DOI 10.1007/978-3-319-39984-3_3

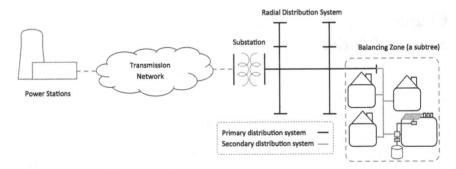

**Fig. 3.1** A schematic diagram of a radial distribution network that emanates from a distribution substation and consists of a number of balancing zones, one of which is illustrated here. Rooftop PV panels, storage systems, and EV chargers are connected to secondary distribution lines, similar to residential and commercial buildings

A balancing zone is defined as a subtree in which reverse flow does not cause any problem for voltage regulators, circuit breakers, and other distribution equipment. Thus, loads can be supplied by any distributed generation resource within the same balancing zone even if this results in reverse flow in some part of the zone. In most distribution systems today, a balancing zone is rooted at a distribution transformer and encompasses the low-voltage residential distribution network fed from a distribution transformer and the loads that are connected to it, as shown in Fig. 3.1. However, in some jurisdictions that have invested in grid modernization, such as in many parts of Germany, power flow is permitted in both directions in the entire distribution network, and therefore, nearly the entire network, including the substation, is contained in a single balancing zone.

### 3.1.1   Network Model

Consider a tree graph $\mathcal{G} = \{\mathcal{B}, \mathcal{L}\}$ that represents the topology of a radial system, comprising a set $\mathcal{B}$ of buses[3] and a set $\mathcal{L}$ of primary distribution lines that connect these buses. Let $\mathcal{B}_z \subset \mathcal{B}$ be the set of buses at the root of each balancing zone, $\mathcal{B}_i$ be the set of buses located downstream of bus $i$, excluding bus $i$ itself, and $\mathcal{L}^i$ be the set of lines located on the unique path from the substation to bus $i$.

To simplify the model, the radial system is studied on a per-phase basis, ignoring the dependency between phases. Additionally, homes, businesses, and other

---

[3] We consider a load bus, laterals radiating from it, and service transformers as a single unit. Thus, downstream loads are aggregated at the load bus as discussed in Sect. 3.2.5.

end-nodes connected to laterals are modelled as single-phase constant complex power loads, i.e., their power consumption is voltage-independent.

A time-slotted model with time slots of equal length $\tau$ (typically of the order of several seconds) is used to study the dynamics of the system. The set of time slots is denoted $\mathcal{T}$ and it is assumed that the network configuration, the demand of inelastic loads, the solar power generated by each panel, the storage output, and the number of plugged-in EVs and their charge power do not change during a time slot. This assumption is necessary to study a dynamical system as a sequence of time slots. To simplify the conversion between energy and power units, *Watt-$\tau$* is used as the unit of energy transmitted, produced, or consumed. For instance, if an EV is charged at the constant rate of 1 Watt in a 1 minute time slot, it consumes 1 W-min of energy.

Consider the distribution system in a time slot $t$. For a bus $i \in \mathcal{B}$, the bus voltage magnitude measured on a per unit basis is denoted $v_i(t)$ and the aggregated real and reactive power consumed at this bus are denoted $p_i(t)$ and $q_i(t)$, respectively.[4] Let bus 0 be the substation (source) bus and $v_0$ be its voltage, which is assumed to be known. The substation bus voltage is used as the base voltage value in the per-unit system; hence, $v_0$ is equal to 1 p.u. here. The impedance of a line connecting bus $i$ to bus $j$ is denoted $z_{ij} = r_{ij} + \mathbf{j}x_{ij}$, where $\mathbf{j}$ is the imaginary unit, and $r_{ij}$ and $x_{ij}$ are the line resistance and reactance, respectively. Furthermore, the sending-end apparent power flow from bus $i$ to bus $j$ is denoted $S_{ij}(t) = P_{ij}(t) + \mathbf{j}Q_{ij}(t)$, where $P_{ij}(t)$ and $Q_{ij}(t)$ are the sending-end real and reactive power flowing between these two buses. Hence, we have $P_{ij}(t) = -P_{ji}(t)$, $Q_{ij}(t) = -Q_{ji}(t)$, and $S_{ij}(t) = -S_{ji}(t)$.

### 3.1.2 Operating Constraints

**Equipment Loading** Resistive heating limits the capability of lines and transformers to transmit power. Hence, every line or transformer in a distribution network has a *nameplate rating* that represents its load carrying capability without overheating.[5] Equipment loading must not exceed its nameplate rating over an extended period of time [8].

An electric utility can specify a control *setpoint*, i.e., a desired loading level, for each line or transformer below its nameplate rating to reduce the risk of equipment overloading by the control system. The aggregate equipment loading, i.e., the sum of elastic and inelastic demands that it supplies, should converge to this setpoint with only a limited number of excursions above the nameplate rating. A conservative utility can ensure a very low congestion level by choosing an appropriately low setpoint. Thus, the setpoints permit the utility to balance utilization and reliability in a distribution network.

---

[4]Note that $p_i$ and $q_i$ are zero if no load is attached to bus $i$.

[5]Line ratings are usually expressed in terms of *ampacity*, whereas transformer ratings are expressed in terms of apparent power.

Let $\xi_{ij}$ be the setpoint associated with a line connecting bus $i$ to $j$ or a transformer installed between bus $i$ and $j$. It is assumed that the setpoints are expressed in Watts for distribution lines and transformers because the electric utility can easily translate line and transformer ratings, which are expressed in Amperes and Volt-amperes, respectively, into the setpoints using a conservative estimate of the power factor and the operating voltage at corresponding nodes. Thus, the following constraint can be written for distribution lines and transformers,[6] including the substation transformer which is located between bus 0 and the next bus:

$$P_{ij}(t) \le \xi_{ij} \qquad \forall\, (i,j) \in \mathcal{L}, t \in \mathcal{T} \tag{3.1}$$

**Voltage Limits** The distribution system code requires the actual service voltage to be maintained within a tolerance band, typically $\pm 5\,\%$ of the nominal voltage [13]. To ensure that the service voltage stays within these strict bounds, electric utilities indirectly control voltage on the primary circuit, taking into account the expected voltage drop along feeders. This involves the control of transformer load tap changers (LTCs), voltage regulators, and switched capacitor banks. The constraint on the bus voltage can be written as:

$$v_{\min}^2 \le v_i(t)^2 \le v_{\max}^2 \qquad \forall\, i \in \mathcal{B}, t \in \mathcal{T} \tag{3.2}$$

where $v_{\min}$ and $v_{\max}$ are the lower and the upper voltage limits that are set to 0.95 p.u. and 1.05 p.u., respectively. Note that this constraint is written in a quadratic form to emphasize that it is linear in $v_i(t)^2$, which appears in (3.15).

**Flow Direction** Reversal of real power flow can negatively impact protection coordination and operation of voltage regulators as distribution circuits are designed with the assumption that the direction of power flow is from the substation to loads at all times. For example, reverse flow conditions can cause *network protectors*, which are installed at distribution transformers, to open unnecessarily and create problems when they reclose [5]. To avoid these problems, many utilities strictly forbid reverse flows outside a balancing zone, meaning that real power cannot be injected to the network at a load bus that represents the root of a balancing zone. This constraint can be written as:

$$p_i(t) \ge 0 \qquad \forall\, i \in \mathcal{B}_{\mathcal{Z}}, t \in \mathcal{T} \tag{3.3}$$

---

[6]We are abusing notation here by referring to the real power supplied by the substation transformer or a service transformer as $P_{ij}$, which also denotes the real power flow at the sending end of a line.

## 3.2 End-Node Models and Constraints

This section presents time-slotted models and operating constraints of inelastic loads, solar inverters, battery storage systems, and EV chargers connected to distribution feeders. These models describe their operation and their state evolution.

### 3.2.1 Inelastic Loads

Inelastic (residential and commercial) loads are connected to the lateral feeders branching from the buses. The real power and reactive power consumed by an inelastic load $i$ in time slot $t$ are denoted $p_i^l(t)$ and $q_i^l(t)$, respectively, and the set of all inelastic loads connected to the distribution network is denoted $\mathcal{I}$. The demand of inelastic loads must be met at all times, unlike the power consumption of elastic loads that can be controlled by the utility within some bounds.

### 3.2.2 Solar Photovoltaic Systems

Consider a rooftop PV system that is connected via a smart inverter to the electrical service panel of a building. This small-scale PV system is single phase and does not need an interconnection transformer. The smart inverter converts the DC output of the system to AC at nominal supply voltage and frequency and provides a wide range of capabilities. These capabilities include injecting or absorbing reactive power and on-demand curtailment of real power[7] [10].

The inverter model adopted here is similar to the models described in [6, 14] and assumes that real and reactive power outputs of an inverter can be controlled independently and simultaneously. Note that inverter losses are ignored here. Let $\mathcal{J}$ denote the set of PV systems in the distribution network, and $\bar{p}_i^s(t)$, $\bar{s}_i^s$, $p_i^s(t)$, and $q_i^s(t)$ denote the available solar power, the inverter's rated apparent power capacity, the real power output, and the reactive power output of the PV system $i$ in a time slot $t$, respectively. Given that the rated apparent power capacity of the inverter is known and the available solar power is measured in this time slot, the following constraints can be written for real and reactive power contributions for this PV system:

$$0 \leq p_i^s(t) \leq \bar{p}_i^s(t) \qquad \forall\, i \in \mathcal{J}, t \in \mathcal{T} \qquad (3.4)$$

$$p_i^s(t)^2 + q_i^s(t)^2 \leq \bar{s}_i^{s2} \qquad \forall\, i \in \mathcal{J}, t \in \mathcal{T} \qquad (3.5)$$

---

[7]Modern inverters can synthesize reactive power just as they produce real power. Despite the fact that the current IEEE 1547 standard for integration of distributed energy resources requires inverters to operate at unity power factor, the use of inverters to assist with voltage regulation is currently an active area of research as they can be controlled on a faster timescale compared to load tap changers and switched capacitors. This possibility is indicated in the proposed IEEE 1547.8 standard.

Note that a negative value of $q_i^s(t)$ means that the inverter is consuming reactive power, while a positive value means that it is injecting reactive power in that time slot. Also note that the second constraint defines a convex set although it is a nonlinear constraint.

### 3.2.3  Battery Storage Systems

Battery storage systems are connected via an interface for AC/DC conversion and a battery management system (BMS) to the electrical service panel of residential and commercial buildings and distribution feeders. The BMS monitors the battery SOC, communicates with external devices, and ensures that charge and discharge operations are within limits of its safe operating area. It is assumed that the storage system can immediately adopt any *feasible* charge or discharge power desired by the controller.

Our battery model, which is similar to one used in [9], assumes that the battery is only capable of absorbing or injecting real power (not reactive power). Let $S$ be the set of battery storage systems in the distribution network and $p_i^b(t)$ be the real power injection of the battery storage system $i$ in time slot $t$. A negative value of $p_i^b(t)$ indicates that the battery is charging in this time slot (acting as a load) and a positive value indicates that it is discharging (acting as a generator). A feasible $p_i^b(t)$ is required to be between the effective maximum charge and discharge powers, denoted $\underline{p}_i^b(t)$ and $\overline{p}_i^b(t)$:

$$-\underline{p}_i^b(t) \leq p_i^b(t) \leq \overline{p}_i^b(t) \quad \forall\, i \in S, t \in T \tag{3.6}$$

In this model, the effective maximum charge and discharge powers of the battery depend on its SOC (a number in [0 1] interval), and the maximum charge and discharge powers supported by the BMS, denoted $\alpha_i^c$ and $\alpha_i^d$. The following two constraints prevent storage from overflowing or underflowing:

$$\underline{p}_i^b(t) = \min\{\alpha_i^c, (\overline{c}_i - c_i(t)) \times \frac{b_i}{\eta_i^c}\}$$

$$\overline{p}_i^b(t) = \min\{\alpha_i^d, (c_i(t) - \underline{c}_i) \times b_i \times \eta_i^d\}$$

where $b_i$ is the energy capacity of the battery, $\eta_i^c$ and $\eta_i^d$ are its charge and discharge efficiencies ($\leq 1$), and $c_i(t)$, $\underline{c}_i$, and $\overline{c}_i$ are its current, minimum, and maximum states of charge ($\in [0\ 1]$), respectively. Hence, the energy content of the battery in time slot $t$ is $c_i(t) \times b_i$.

The state of charge evolution of the battery can be written as:

$$
c_i(t) = \begin{cases} c_i(t-1) - \eta_i^c \times \frac{p_i^b(t-1)}{b_i} & \text{if } -\underline{p}_i^b(t-1) \le p_i^b(t-1) \le 0 \\ c_i(t-1) - \frac{p_i^b(t-1)}{\eta_i^d \times b_i} & \text{if } \qquad 0 < p_i^b(t-1) \le \overline{p}_i^b(t-1) \end{cases}
$$

### 3.2.4  Electric Vehicle Chargers

Smart EV chargers connect to the electric circuit of residential and commercial buildings. It is assumed that chargers only consume real power and EV batteries cannot be discharged to offer system services as in the V2G case; the energy stored in the EV battery is solely used by the motor to drive the vehicle. This is the main difference between EVs and dedicated storage systems; the other difference being that EVs can drive away, unlike stationary storage systems.

A smart charger is called *active* when an EV is plugged in and ready to be charged. An active smart charger can provide any feasible charge power desired by the operator. The charge power is assumed to be independent of the SOC of the connected EV.[8] Let $\mathcal{E}$ be the set of EV chargers connected to the distribution network. The charging load of an EV $i$ is characterized by its maximum and minimum demands in a given time slot $t$, which are denoted $\overline{p}_i^e(t)$ and $\underline{p}_i^e(t)$ and are defined as:

$$
\overline{p}_i^e(t) = \min\{\beta_i, e_i(t)\}
$$
$$
\underline{p}_i^e(t) = \min\{\overline{p}_i^e(t), \frac{e_i(t)}{d_i}\}
$$

where $e_i(t)$ is the amount of energy required to fill the battery,[9] $\beta_i$ is the maximum charge power supported by the charger, and $d_i$ is the charging *deadline* of the EV expressed in number of time slots. Hence, a feasible charging rate for this time slot, denoted $p_i^e(t)$, must be between the maximum and minimum demands:

$$
\underline{p}_i^e(t) \le p_i^e(t) \le \overline{p}_i^e(t) \qquad \forall\, i \in \mathcal{E}, t \in \mathcal{T} \tag{3.7}
$$

In this work the minimum demand of a charger, $\underline{p}_i^e$, is set to zero[10] since the proposed charging scheme is *best-effort* and does not guarantee to fulfill the charging demand before the deadline (it might be infeasible to meet the deadline).

---

[8]This is a simplification. In fact, when the SOC is high, charge power must be limited to prevent overvoltage.

[9]A charger $i$ sets $e_i(t)$ to zero if it is inactive at the beginning of time slot $t$. Hence, $\underline{p}_i^e(t) = \overline{p}_i^e(t) = 0$ in that time slot.

[10]This is similar to the case that $d_i = \infty$ for every charger $i$.

Finally, the amount of energy required to fill the battery evolves according to the following equation:

$$e_i(t) = e_i(t-1) - \gamma_i^c \times p_i^e(t-1)$$

where $\gamma_i^c$ is the charge efficiency of the battery.

### 3.2.5   Load Aggregation at Buses

Given the models of loads and active end-nodes, it is straightforward to derive the total real and reactive power consumed at each bus. Let $\mathbf{A}^l$, $\mathbf{A}^e$, $\mathbf{A}^s$, and $\mathbf{A}^b$ encode the point of connection of inelastic loads, EV chargers, PV systems, and battery storage systems. For example, $\mathbf{A}_{ij}^l$ is 1, if an inelastic load indexed by $i$ is connected under a load bus $j$, and is 0 otherwise. The other matrices are defined in a similar way. Thus, the total real and reactive power consumed at bus $j$ in time slot $t$ can be obtained as follows:

$$p_j(t) = \sum_{i:\mathbf{A}_{ij}^l=1} p_i^l(t) + \sum_{i:\mathbf{A}_{ij}^e=1} p_i^e(t) - \sum_{i:\mathbf{A}_{ij}^s=1} p_i^s(t) - \sum_{i:\mathbf{A}_{ij}^b=1} p_i^b(t) \qquad (3.8)$$

$$q_j(t) = \sum_{i:\mathbf{A}_{ij}^l=1} q_i^l(t) - \sum_{i:\mathbf{A}_{ij}^s=1} q_i^s(t) - q_j^c(t) \qquad \forall j \in \mathcal{B}, t \in \mathcal{T} \qquad (3.9)$$

where $q_j^c(t)$ represents the total reactive power provided by shunt capacitors connected to bus $j$ in time slot $t$. Hence, $q_j^c$ is zero when no shunt capacitor is connected to a bus. Note that battery storage systems and EV chargers are assumed to operate at unity power factor.

## 3.3   Power Flow Model

Power flow in a balanced radial distribution system can be approximated with single-phase recursive branch flow equations, known as *DistFlow* equations [2–4]. This specific formulation leads to efficient solution methods for computing bus voltages and branch flows, given the real and reactive power drawn from or injected to every load bus. This section presents the DistFlow model and a linearized power flow model based on an approximation that ignores power losses.

The DistFlow model can be described with the following equations:

$$P_{ij}(t) = p_j(t) + \sum_{k \neq i:(j,k) \in \mathcal{L}} P_{jk}(t) + r_{ij} \frac{P_{ij}(t)^2 + Q_{ij}(t)^2}{v_i(t)^2} \qquad (3.10)$$

$$Q_{ij}(t) = q_j(t) + \sum_{k \neq i:(j,k) \in \mathcal{L}} Q_{jk}(t) + x_{ij} \frac{P_{ij}(t)^2 + Q_{ij}(t)^2}{v_i(t)^2} \qquad (3.11)$$

$$v_j(t)^2 = v_i(t)^2 - 2(r_{ij}(t)P_{ij}(t) + x_{ij}(t)Q_{ij}(t)) + (r_{ij}^2 + x_{ij}^2) \frac{P_{ij}(t)^2 + Q_{ij}(t)^2}{v_i(t)^2} \qquad (3.12)$$

where $\frac{P_{ij}(t)^2 + Q_{ij}(t)^2}{v_i(t)^2}$ is the square of the current magnitude that is being carried by the line connecting bus $i$ to bus $j$, meaning that the quadratic terms in the above equations represent line losses. Note that an OPF problem that incorporates the DistFlow model is not convex and, therefore, finding its solution(s) will be of exponential complexity in the number of nodes.

Since losses are typically quite smaller than the real and reactive power flow components, an approximation that ignores the higher order loss terms introduces only a small error on the order of 1 %. This approximate power flow model is referred to as the *simplified DistFlow*. This model was originally proposed in [4] and has been used several times to formulate convex optimal control problems for distribution networks, see for example [1, 7, 14]. The simplified DistFlow equations can be written as follows after unfolding the recursions:

$$P_{ij}(t) = \sum_{k \in \mathcal{B}_i} p_k(t) \qquad \forall\, (i,j) \in \mathcal{L}, t \in \mathcal{T} \qquad (3.13)$$

$$Q_{ij}(t) = \sum_{k \in \mathcal{B}_i} q_k(t) \qquad \forall\, (i,j) \in \mathcal{L}, t \in \mathcal{T} \qquad (3.14)$$

$$v_j(t)^2 = v_i(t)^2 - 2(r_{ij}P_{ij}(t) + x_{ij}Q_{ij}(t)) \qquad \forall\, j \in \mathcal{B}, t \in \mathcal{T}$$

$$= v_0^2 - 2 \left( \sum_{k \in \mathcal{B}} p_k(t) \sum_{(m,n) \in \mathcal{L}^j \cap \mathcal{L}^k} r_{mn} + \sum_{k \in \mathcal{B}} q_k(t) \sum_{(m,n) \in \mathcal{L}^j \cap \mathcal{L}^k} x_{mn} \right) \qquad (3.15)$$

where $\mathcal{B}_i$ is the set of buses downstream of bus $i$ and $\mathcal{L}^j \cap \mathcal{L}^k$ is the set of lines that supply both bus $j$ and bus $k$. Observe that these equations are linear in the squared voltage magnitudes, and real and reactive power flows. Also remark that the linear branch flow equations (3.13)–(3.15) make it possible to enforce capacity and voltage limits in optimal control problems without losing computational tractability.

## 3.4  Fairness and Resource Allocation

It is crucial for an allocation policy to ensure that users are treated *fairly* and no user is starved of service in a system with constrained resources that are shared by many users. Fairness can be defined in different ways depending on the context. This has motivated the development of an optimization framework to unify various fairness criteria for systems with single or multiple types of resources.

The idea is to attribute a *utility*, i.e., a measure of satisfaction, to every user, assuming that users are greedy (in terms of the resources they want) and their utility increases with the amount of resources allocated to them in an interval that spans over one or several time slots. A proportionally fair allocation is an allocation that maximizes a global objective function defined as the sum of the logarithm of the utility function of all users [11].

A fair allocation is a socially optimal allocation, i.e., an allocation that maximizes a utilitarian criterion which is a function of the utilities of individuals and can be defined in many different ways. There are several well-established axiomatically justified notions of fairness, such as max-min fairness, proportional fairness, minimum potential delay fairness, and the more general notion of utility proportional fairness; these notions of fairness differ in the choice of the global objective function. This work adopts the notion of proportional fairness since it is the only one that provides a scale invariant Pareto optimal solution, which is consistent with axioms of fairness in game theory [16].

In Chap. 4, the notion of fair resource allocation is extended to power distribution systems with a certain population of controlled loads, such as EV chargers. The goal is to allocate the total available real power to EV chargers in a fair and efficient manner. A utility is attributed to each EV owner, which is defined as the instantaneous charge power of their EV.

## 3.5  Chapter Summary

In this chapter, simplified time-slotted models are presented for conventional loads and the active end-nodes, and an approximate linear branch flow model is introduced for radial distribution systems. These branch flow equations can be incorporated in the formulation of convex optimal control problems, as discussed in the next chapter. Finally, the notion of proportional fairness is described in the context of the allocation of real power to EV chargers in the distribution system.

# References

1. Arnold DB, Negrete-Pincetic M, Stewart EM, Auslander DM, Callaway DS (2015) Extremum Seeking control of smart inverters for VAR compensation. In: PES general meeting. IEEE, pp 1–5
2. Baran M, Wu F (1989) Network reconfiguration in distribution systems for loss reduction and load balancing. IEEE Trans Power Deliv 4(2):1401–1407
3. Baran M, Wu F (1989) Optimal capacitor placement on radial distribution systems. IEEE Trans Power Deliv 4(1):725–734
4. Baran M, Wu F (1989) Optimal sizing of capacitors placed on a radial distribution system. IEEE Trans Power Deliv 4(1):735–743
5. Coddington M, Kroposki B, Basso T, Lynn K, Sammon D, Vaziri M, Yohn T (2009) Photovoltaic systems interconnected onto secondary network distribution systems-success stories. Report NREL/TP-550-45061, National Renewable Energy Laboratory
6. Farivar M, Neal R, Clarke C, Low S (2012) Optimal inverter VAR control in distribution systems with high PV penetration. In: IEEE PES general meeting, pp 1–7
7. Farivar M, Chen L, Low S (2013) Equilibrium and dynamics of local voltage control in distribution systems. In: IEEE conference on decision and control, pp 4329–4334
8. Featheringill WE (1983) Power transformer loading. IEEE Trans Appl Ind IA-19(1):21–27
9. Ghiassi-Farrokhfal Y, Kazhamiaka F, Rosenberg C, Keshav S (2015) Optimal design of solar pv farms with storage. IEEE Trans Sustain Energy 6(4):1586–1593
10. Katiraei F, Sun C, Enayati B (2015) No inverter left behind: Protection, controls, and testing for high penetrations of PV inverters on distribution systems. IEEE Power Energy Mag 13(2): 43–49
11. Kelly F (1997) Charging and rate control for elastic traffic. Eur Trans Telecommun 8(1):33–37
12. Kersting W (2012) Distribution system modeling and analysis, 3rd edn. Taylor & Francis, Boca Raton
13. Meier A (2006) Electric power systems: a conceptual introduction. Wiley-IEEE Press, New York
14. Turitsyn K, Sulc P, Backhaus S, Chertkov M (2010) Distributed control of reactive power flow in a radial distribution circuit with high photovoltaic penetration. In: IEEE PES general meeting, pp 1–6
15. Walling R, Saint R, Dugan R, Burke J, Kojovic L (2008) Summary of distributed resources impact on power delivery systems. IEEE Trans Power Deliv 23(3):1636–1644
16. Yaïche H, Mazumdar R, Rosenberg C (2000) A game theoretic framework for bandwidth allocation and pricing in broadband networks. IEEE/ACM Trans Netw 8(5):667–678

# Chapter 4
# Optimal Control of Active End-Nodes

**Abstract** This chapter studies a radial distribution system, which is divided into a number of balancing zones, and proposes a decentralized scheme for the joint control of EV chargers, PV inverters, and storage systems that are under the exclusive control of the utility. The proposed open-loop control scheme exploits the synergy between EV chargers and PV inverters to cancel out their effects on distribution circuits, and relies on a sophisticated distribution system model and near real-time measurements of the end-nodes to simultaneously achieve the utility-defined objectives. Our decentralized control scheme is compared to two conservative, fully distributed control schemes that enable customers to control the active end-nodes installed in their premises without the benefit of coordination from the utility.

## 4.1 The Synergy Between EV Chargers and PV Inverters

Many utilities have begun to experience the impacts of a high concentration of PV systems and an increasing number of EV chargers on their distribution circuits. As discussed in Chap. 2, overvoltage and undervoltage conditions, transformer and feeder overloads, and reverse power flow are more likely to happen in these distribution networks. Reverse flow, which occurs when solar generation exceeds feeder loading, could cause protection coordination problems and overuse of voltage regulators and switched capacitors, shortening their expected life cycle [1].

To mitigate these problems, utilities can limit PV and EV charger installations in size and number; but this comes at the price of a significant reduction in the efficiency of the grid and the flexibility that it offers to its customers. A more promising approach would be to exploit the synergy between EV chargers, storage systems, and PV inverters to reliably accommodate a higher penetration of these active end-nodes in existing distribution systems. For example, the charge power of EV chargers and storage systems located in a balancing zone can be controlled to absorb solar generation locally when it peaks. Similarly, real and reactive power outputs of PV inverters can be adjusted to match demands of EV chargers within the same balancing zone. This enhances reliability, enables charging a larger population of EVs, reduces

© The Author(s) 2016

O. Ardakanian et al., *Integration of Renewable Generation and Elastic Loads into Distribution Grids*, SpringerBriefs in Electrical and Computer Engineering, DOI 10.1007/978-3-319-39984-3_4

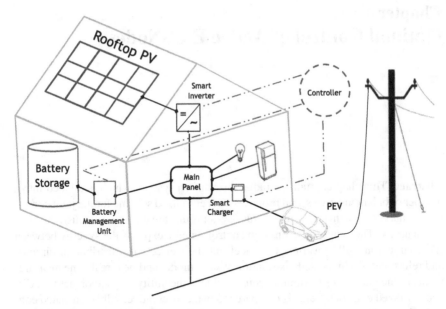

**Fig. 4.1** A schematic diagram of a small business with a rooftop PV system, a battery storage system, a PEV, and other inelastic loads that are connected to the mains via an electrical service panel. The smart inverter, the smart EV charger, and the battery management system communicate with the upstream controller(s) over a broadband communication network, depicted by *dashed lines*

wasteful and expensive solar generation curtailment and overall carbon emissions, and most importantly eliminates the trade-off between reliability and efficiency.

Consider a radial distribution system that supplies homes and small businesses constituting a number of balancing zones. The active end-nodes, including solar inverters, storage systems, and EV chargers, are assumed to be installed at small businesses[1] and EVs are assumed to be parked and connected at these small businesses during business hours. Hence, the chargers are likely to be active during the day when solar energy can be harnessed. The active end-nodes connect to the electrical service panel of the building and communicate with an upstream controller, which will be discussed in Sect. 4.2, over a proprietary network as illustrated in Fig. 4.1. Assuming that the utility is granted remote control and monitoring of active end-nodes[2] and pays for solar generation even if it is curtailed, we design an optimal control scheme for EV chargers, PV inverters, and storage systems to simultaneously achieve multiple utility-defined objectives,subject to the network and end-node constraints described

---

[1]A business does not necessarily install all three technologies.

[2]Customers may relinquish control of active end-nodes in exchange for a fixed reduced electricity price. In this case, any control signal issued by the utility is assured of an immediate cooperative response.

in Sects. 3.1.2 and 3.2. This scheme enables sharing of solar generation and stored energy within each balancing zone.

The utility has four kinds of control knobs in the last mile of the distribution network, namely the charge power of EV chargers, $p^e(t)$, the real and reactive power outputs of inverters, $p^s(t)$ and $q^s(t)$, and the real power contribution of storage systems, $p^b(t)$. The optimal control is found by solving a sequence of two optimization problems for every time slot in a decentralized fashion (at the level of the substation and the level of balancing zones), where the length of each time slot is 1 min during which the number of active end-nodes and household and business demands are assumed to be constant. The proposed control is *myopic* as the objective functions depend only on the charge power of EV chargers, real and reactive outputs of inverters, and storage operations in the current time slot, ignoring their future and past dynamics. The myopic approach is reasonable given that EVs are unpredictable and can drive off at any time.

## 4.2 Control Objectives

The utility must meet the demand of homes and businesses at all times. Additionally, it seeks to operate active end-nodes so as to maximize its revenue, assuming that it has full control over EV chargers, PV inverters, and battery management systems. The utility is also required to implement government mandates, such as expanding renewable energy generation and cutting emissions. This leads to a multi-objective optimization problem that can be solved to obtain the optimal control.

These objectives are conflicting, so any controller design will need to make a trade-off between the objectives. Our approach is to put the objectives into a total ordering, as described next. Note that a different ordering would result in a different control system. The control objectives that we consider are listed below in descending order of importance to the utility: (1) maximize the utility's revenue by maximizing the total power delivered to elastic loads from different sources and, in particular, by allocating the available power to connected EVs in a fair manner, (2) minimize the curtailment of solar power, (3) minimize the use of conventional power from the grid, thereby reducing carbon emissions. The following sections discuss these control objectives and argue that this particular ordering is both reasonable and necessary. Section 4.3 then formulates a series of two optimization problems to achieve these objectives in the order specified above.

Note that we did not take into account other plausible objectives such as minimizing energy losses, minimizing the peak-to-average ratio, or minimizing the amount of storage needed in the system. These objectives would form a fruitful avenue for future work.

### 4.2.1   Objective 1—Maximizing Revenue Through Fair Power Allocation to EV Chargers

We believe that the primary objective of the electric utility will always be to maximize its revenue.[3] Assuming that the revenue is a strictly monotone function of the supplied power, maximizing the revenue is the same as maximizing the total supplied power. Since the demand of inelastic loads must be met at all times, a revenue-maximizing strategy is the one that maximizes the total real power allocated to elastic loads in every time slot.[4]

There are possibly many feasible revenue-maximizing power allocations in every time slot, since real power can be distributed among active chargers in different ways, all having the same total use of real power. We prefer the allocation that is fair to the connected EVs. As discussed in Sect. 3.4, it can be assumed that EV owners are greedy and want to finish charging their EVs as soon as possible; therefore, at time $t$, the utility attributed to the EV owner $i$ is equal to the charge power currently adopted by its charger, $p_i^e(t)$.

A global optimization problem is formulated to maximize the sum of the logarithm of the utility function for EV owners. This choice of the objective function guarantees that real power is allocated in a proportionally fair manner among active EV chargers. Note that the logarithm of the utility function of each user, i.e., $\log(p_i^e(t))$, is an infinitely differentiable, increasing, and strictly concave function in its domain, and therefore, the global objective function is also concave. Also note that the proportionally fair allocation is indeed a revenue-maximizing allocation. This is an appealing property of proportional fairness in that it utilizes all available resources.

### 4.2.2   Objective 2—Minimizing Solar Curtailment

Curtailing solar generation is a forfeiture of inexpensive green energy. This motivates our choice of minimizing the curtailment of distributed solar generation, which is equivalent to maximizing the use of solar power, as the secondary objective of the electric utility. Even when solar generation exceeds the aggregate demand of a balancing zone the excess energy can be used to charge storage systems within the same balancing zone. Nevertheless, curtailment cannot be avoided at all times; excess solar generation must be curtailed when it cannot be stored or exported due to the constraints presented in Chap. 3. Smart inverters are capable of curtailing solar generation in these occasions.

---

[3] Recall that it is assumed that the utility pays for solar generation even if it is curtailed. Thus, its revenue only depends on the amount of energy delivered to the customers.

[4] We do not take into account the energy that can be charged into storage systems when defining the revenue-maximizing control strategy for a time slot. This is because this energy is not actually consumed and will be used at some point to supply loads (with some losses). Hence, the utility does not increase its revenue in the long run by storing energy in the distribution network.

Note that there are, in general, many possible ways, i.e., many combinations of conventional, solar, and stored powers, to deliver the computed maximum supplied power. This objective forces the selection of the one that uses as much solar power as available. Hence, it is not redundant given the revenue maximization objective.

### 4.2.3 Objective 3—Minimizing the Use of Conventional Power

Displacing conventional power supplied by the substation with solar power produced instantaneously by rooftop PV systems or stored in battery storage systems in previous time slots reduces the overall cost and carbon emissions of electricity generation as well as transmission losses. Hence, the utility would strive to minimize the use of conventional power to improve the power system efficiency, reduce transmission losses, and comply with external mandates. The use of conventional power is therefore restricted to when household and business demands cannot be met entirely by PV and storage systems.

Note that this objective is not redundant given the first two objectives because conventional power can be displaced with discharged power from storage systems without having any impact on the first two objectives.

## 4.3 Optimal Control

This section describes a series of two optimization problems that generate the optimal control in every time slot, and discusses how the second problem can be decomposed into a number of decoupled problems. A decentralized control scheme that solves these optimization problems at two different levels is proposed in the next section.

### 4.3.1 Optimization Problems

A multi-step optimization is required to satisfy the three objectives specified in Sect. 4.2 without using an arbitrary scalarization, i.e., a weighted sum of the objectives. In particular, reducing the use of conventional power is in conflict with the revenue-maximization objective because it can reduce the total supplied power; therefore, a two-step optimization is inevitable. These two optimization problems are discussed next.

### 4.3.1.1  Revenue-Maximizing Fair Allocation with Minimum Solar Curtailment

The first optimization problem aims at minimizing the solar curtailment and maximizing the revenue, while being fair to the active chargers. Since the first two objective functions of Sect. 4.2 are not conflicting, it is possible to optimize them at the same time without introducing weight terms. Specifically, increasing the use of solar power does not negatively impact the optimal power allocation to EV chargers. Hence, the optimizer of the sum of these two objectives is the solution to *any* weighted sum of these two objectives.[5]

Assuming that the impedance of the main feeders, real and reactive power consumption of homes and businesses, the setpoint of feeders and transformers, the available solar power at the point of connection of PV systems, and the set of active end-nodes and their parameters are known in the beginning of every time slot, Problem 1 can be posed as a nonlinear optimization problem, where the control variables are $\mathbf{p}^e(t), \mathbf{p}^b(t), \mathbf{p}^s(t), \mathbf{q}^s(t)$.

Problem 1: the global power allocation problem
**Inputs:** $\mathbf{p}^l(t), \mathbf{q}^l(t), \xi, \overline{\mathbf{p}}^s(t), \overline{\mathbf{s}}^s, \overline{\mathbf{p}}^e(t), \overline{\mathbf{p}}^b(t), \underline{\mathbf{p}}^b(t), \mathcal{I}, \mathcal{E}, \mathcal{J}, \mathcal{S}$

$$\max_{\mathbf{p}^e(t), \mathbf{p}^b(t), \mathbf{p}^s(t), \mathbf{q}^s(t)} \sum_{i \in \mathcal{E}} \log(p_i^e(t)) + \sum_{i \in \mathcal{J}} p_i^s(t) \tag{4.1}$$

subject to

End-node Constraints $(3.4 - 3.7)$
System Constraints $(3.1 - 3.3)$
Bus Injection Equations $(3.8 - 3.9)$
Power Flow Equations $(3.13 - 3.15)$

Problem 1 is subject to the linearized power flow equations, the real and reactive power injection equations for buses, the distribution systems constraints, and the end-node constraints. Note that this nonlinear optimization problem is convex as it maximizes a concave function which is the sum of two concave functions, one linear and one nonlinear, subject to affine equality constraints and linear and quadratic inequality constraints[6] that define a convex set. Therefore, it has a unique solution. The unique proportionally fair power allocation to EV chargers in time slot $t$ is represented by $\tilde{\mathbf{p}}^e(t)$, and the optimal real and reactive power contributions of PV inverters are represented by $\tilde{\mathbf{p}}^s(t)$ and $\tilde{\mathbf{q}}^s(t)$, respectively. Here the upright boldface letters represent vectors.

---

[5]Nevertheless, algorithmically weight terms are important because they influence how fast the optimal solution is found.

[6]The quadratic constraints pertain to the apparent power capacity of solar inverters.

### 4.3.1.2 Minimizing the Use of Conventional Power

Given $\tilde{\mathbf{p}}^e(t)$, $\tilde{\mathbf{p}}^s(t)$, and $\tilde{\mathbf{q}}^s(t)$, the second optimization problem, called Problem 2, aims at minimizing the power supplied by the grid in a time slot, which can be written as:

$$P_{\text{grid}}(t) = \sum_{i \in \mathcal{I}} p_i^l(t) + \sum_{i \in \mathcal{E}} \tilde{p}_i^e(t) - \sum_{i \in \mathcal{J}} \tilde{p}_i^s(t) - \sum_{i \in \mathcal{S}} p_i^b(t)$$

Since the three first terms in the right hand side of this equation are fixed, maximizing the total power discharged from storage systems minimizes the use of conventional power supplied by the grid. Given real and reactive power consumption of homes and businesses, the setpoints of feeders and transformers, the solution to the first optimization problem, the available solar power at the point of connection of PVs in each time slot, the set of active end-nodes, and their parameters, Problem 2 is posed to determine the optimal control of storage systems. This problem includes only the end-node constraints that pertain to storage systems (Constraints 4.3 and 4.4) as the operations of other active end-nodes have been determined already.

As a practical matter, all storage systems located in the same balancing zone must be either charging or discharging in a given time slot; otherwise, control may discharge one storage system and use the energy stored in that system to charge another storage system in the same zone. This would be neutral in terms of the objective function but would affect the amount of energy that can be discharged from the storage systems in the future time slots. Particularly, energy transfer between storage systems that are within the same zone results in waste of energy due to storage charge and discharge inefficiencies. To rule out such controls, all storage systems located in the same zone are forced to either charge or discharge in each time slot, thereby maximizing the system efficiency implicitly.

Problem 2: the global storage problem

**Inputs:** $\mathbf{p}^l(t), \mathbf{q}^l(t), \xi, \tilde{\mathbf{p}}^s(t), \tilde{\mathbf{q}}^s(t), \tilde{\mathbf{p}}^e(t), \overline{\mathbf{p}}^b(t), \underline{\mathbf{p}}^b(t), \mathcal{I}, \mathcal{E}, \mathcal{J}, \mathcal{S}$

$$\max_{\mathbf{p}^b(t)} \sum_{i \in \mathcal{S}} p_i^b(t) \tag{4.2}$$

subject to

$$0 \leq p_i^b(t) \leq \overline{p}_i^b(t) \qquad i \in \mathcal{S}^D \tag{4.3}$$

$$-\underline{p}_i^b(t) \leq p_i^b(t) \leq 0 \qquad i \in \mathcal{S}^C \tag{4.4}$$

System Constraints $(3.1 - 3.3)$

Bus Injection Equations $(3.8 - 3.9)$

Power Flow Equations $(3.13 - 3.15)$

Let us denote the set of storage systems that must be charged and the set of storage systems that must be discharged by $\mathcal{S}^C$ and $\mathcal{S}^D$, respectively, which are defined as:

$$\mathcal{S}^C = \left\{ i \in \mathcal{S} | \mathbf{A}^b_{ij} = 1, j \in \mathcal{B}^C \right\} \tag{4.5}$$

$$\mathcal{S}^D = \left\{ i \in \mathcal{S} | \mathbf{A}^b_{ij} = 1, j \in \mathcal{B}^D \right\} \tag{4.6}$$

where $\mathcal{B}^C$ and $\mathcal{B}^D$ are balancing zones in which every storage system must be charged and discharged, respectively. These two sets are defined as:

$$\mathcal{B}^C = \left\{ j \in \mathcal{B}_Z | \sum_{i:A^s_{ij}=1} \tilde{p}^s_i(t) > \sum_{i:A^l_{ij}=1} p^l_i(t) + \sum_{i:A^e_{ij}=1} \tilde{p}^e_i(t) \right\} \tag{4.7}$$

$$\mathcal{B}^D = \left\{ j \in \mathcal{B}_Z | \sum_{i:A^s_{ij}=1} \tilde{p}^s_i(t) \le \sum_{i:A^l_{ij}=1} p^l_i(t) + \sum_{i:A^e_{ij}=1} \tilde{p}^e_i(t) \right\} \tag{4.8}$$

Problem 2 can have multiple solutions, each minimizing the use of conventional power. An optimal control for storage systems in time slot $t$ is denoted $\tilde{\mathbf{p}}^b(t)$.

Observe that Problem 2 is separable because no constraint couples storage systems that belong to two different balancing zones.[7] Thus, this problem can be decomposed into smaller subproblems of the forms (4.9) and (4.10) for "charging" and "discharging" balancing zones, respectively. These subproblems are LP. Solving each of these subproblems can be delegated to a controller installed at the edge of the corresponding balancing zone as discussed in the next section.

---

**Problem 2-1: the charging balancing zone problem:** $j \in \mathcal{B}^C$
**Inputs:** $\mathbf{p}^l(t), \mathbf{q}^l(t), \xi, \tilde{\mathbf{p}}^s(t), \tilde{\mathbf{q}}^s(t), \tilde{\mathbf{p}}^e(t), \overline{\mathbf{p}}^b(t), \underline{\mathbf{p}}^b(t), \mathcal{I}, \mathcal{E}, \mathcal{J}, \mathcal{S}$

---

$$\max_{\mathbf{p}^b(t)} \sum_{i \in \mathcal{S}} p^b_i(t) \tag{4.9}$$

subject to

$$-\underline{p}^b_i(t) \le p^b_i(t) \le 0 \qquad i \in \mathcal{S}^C$$

System Constraints $(3.1 - 3.3)$
Bus Injection Equations $(3.8 - 3.9)$
Power Flow Equations $(3.13 - 3.15)$

---

[7]Line and transformer capacity constraints that are outside balancing zones can be ignored in Problem 2. This is because storage systems are not charged from the grid due to the third objective and their optimal control, i.e., the solution of Problem 2, does not overload any line or transformer if the capacity constraints are ignored because Problem 1 had a feasible solution.

Problem 2-2: the discharging balancing zone problem: $j \in \mathcal{B}^D$
**Inputs:** $\mathbf{p}^l(t), \mathbf{q}^l(t), \xi, \tilde{\mathbf{p}}^s(t), \tilde{\mathbf{q}}^s(t), \tilde{\mathbf{p}}^e(t), \overline{\mathbf{p}}^b(t), \underline{\mathbf{p}}^b(t), \mathcal{I}, \mathcal{E}, \mathcal{J}, \mathcal{S}$

$$\max_{\mathbf{p}^b(t)} \sum_{i \in \mathcal{S}} p_i^b(t) \tag{4.10}$$

subject to

$$0 \leq p_i^b(t) \leq \overline{p}_i^b(t) \qquad i \in \mathcal{S}^D$$

System Constraints $(3.1 - 3.3)$
Bus Injection Equations $(3.8 - 3.9)$
Power Flow Equations $(3.13 - 3.15)$

## 4.4  Multi-Tier Control Architecture

An electric utility may control thousands of PV panels, storage systems, and EV chargers. Critical to the control scheme are the measurements that are used as input to the optimization problems. Getting these measurements requires a measurement infrastructure that can be combined with the control infrastructure. This calls for the design of an overall architecture that enables scalable, robust, timely, and secure data transfer between measurement and control nodes. To this end, a multi-tier control architecture that consists of a centralized substation controller that coordinates control with a set of controllers corresponding to balancing zones is adopted here.

A reliable communication network connects the substation controller to the balancing zone controllers and to measurement devices installed at all homes and businesses. These devices measure residential and commercial demands and the parameters of the active end-nodes, and send them periodically (once every time slot of length 1 min) to their upstream controller as illustrated in Fig. 4.2. Specifically, each balancing zone controller receives near real-time measurements (i.e., with a delay much smaller than 1 second) of the real and reactive power consumption of inelastic loads, the maximum demand of active EV chargers, the available real power at PV systems, and the maximum feasible charge and discharge powers of storage systems from downstream active end-nodes. The controllers treat these measurements as estimates of the corresponding values in the next time slot.

**Decentralized Algorithm** Control actions are computed jointly by the substation controller and balancing zone controllers as follows:

Step 1: active end-nodes communicate their latest measurements to the substation controller via their zone controller

Step 2: the substation controller runs Algorithm 1.

**Fig. 4.2** A schematic of the
multi-tier control
architecture showcasing the
substation controller and two
balancing zones with their
controller and measurement
devices installed at the
end-nodes. Communication
links between measurement
nodes and upstream
controllers are depicted by
*dashed arrows*

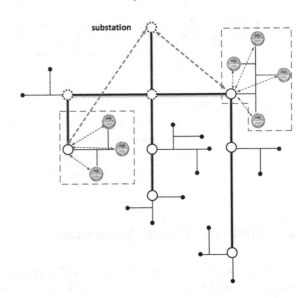

Step 3: every balancing zone controller runs Algorithm 2 upon receiving control
decisions from the substation controller.

Step 4: every active end-node carries out the optimal control received from its
upstream zone controller in the beginning of the next time slot.

---

**Algorithm 1**: Algorithm run by the substation controller

---

**Data**: $\xi_{ij}$, $v_{min}$, $v_{max}$, $\mathcal{B}_{\mathcal{Z}}$, $z_{ij}$, $\bar{\mathbf{s}}^s$

**while** *true* **do**

    Receive recent measurements of $\mathbf{p}^l$, $\mathbf{q}^l$, $\overline{\mathbf{p}}^s$, $\overline{\mathbf{p}}^e$, $\overline{\mathbf{p}}^b$, $\underline{\mathbf{p}}^b$ from end-nodes;

    Estimate $\mathbf{p}^l$, $\mathbf{q}^l$, $\overline{\mathbf{p}}^s$, $\overline{\mathbf{p}}^e$, $\overline{\mathbf{p}}^b$, $\underline{\mathbf{p}}^b$ for the next time slot;

    Solve Problem 1 for the next time slot;

    Send $\tilde{\mathbf{p}}^e$, $\tilde{\mathbf{p}}^s$, $\tilde{\mathbf{q}}^s$ to downstream controllers of $\mathcal{B}_{\mathcal{Z}}$;

    Wait until the next **clock tick**;

**end**

---

## 4.5   Benchmarks

To compare the performance of the decentralized control scheme with schemes that
are already used in the field, two fully distributed control schemes (one that utilizes
local storage and one that does not) are used as benchmarks. These controllers run
at individual businesses to control the operation of active end-nodesthat are installed

---

**Algorithm 2**: Algorithm run by a zone controller

---

**Data:** $\xi_{ij}$, $v_{min}$, $v_{max}$, $z_{ij}$, $\bar{\mathbf{s}}^s$

**while** *true* **do**

    Receive recent measurements of $\mathbf{p}^l$, $\mathbf{q}^l$, $\overline{\mathbf{p}}^b$, $\mathbf{p}^b$ from end-nodes;

    Receive computed $\tilde{\mathbf{p}}^e$, $\tilde{\mathbf{p}}^s$, $\tilde{\mathbf{q}}^s$ from substation controller;

    Estimate $\mathbf{p}^l$, $\mathbf{q}^l$, $\overline{\mathbf{p}}^b$, $\underline{\mathbf{p}}^b$ for the next time slot;

    **if** *zone* $\in \mathcal{B}^C$ **then**

        | Solve Problem 2-1 for the next time slot;

    **else**

        | Solve Problem 2-2 for the next time slot;

    **end**

    Send $\tilde{\mathbf{p}}^e$, $\tilde{\mathbf{p}}^s$, $\tilde{\mathbf{q}}^s$, $\tilde{\mathbf{p}}^b$ to downstream end-nodes;

    Wait until the next **clock tick**;

**end**

---

there (i.e., local resources), using local measurements only. Both schemes aim to limit the output of solar inverters to meeting the local demand; thus, they curtail solar generation when it exceeds the aggregate local demand and do not allow sharing within the balancing zones. These schemes cannot control the reactive power output of inverters or adjust the charge power of EV chargers because they are not aware of the distribution network model and also cannot observe voltage and power flow at upstream buses. Thus, voltage and congestion problems are still possible due to the operation of EV chargers. Nevertheless, these schemes serve as benchmarks for the proposed control scheme as discussed in the next chapter.

In some jurisdictions, less conservative schemes can be used to control the output of inverters. These schemes permit the export of excess solar generation to the grid as long as the voltage level at inverters' point of connection stays within some bound. These control schemes are not used as our benchmarks since calculating the voltage at the point of connection of inverters requires the knowledge of the impedance of secondary distribution lines, which was not available to us as academic researchers.

### 4.5.1 Without Local Storage

The first scheme assumes that small businesses do not have dedicated storage systems, and therefore, solar generation must be curtailed when it exceeds the aggregate local demand, which is the sum of the demand of the small business and the maximum demand of the EV charger installed at the small business. Each controller aims at (1) charging the EV at the maximum rate, and (2) minimizing solar curtailment from the local PV system, subject to the constraint that there is no export of real power to the grid. Thus, the controller implements the following rules in the given order:

$$p_i^e(t) = \overline{p}_i^e(t) \tag{4.11}$$

$$p_i^s(t) = \min\{\overline{p}_i^s(t), p_i^l(t) + p_i^e(t)\} \tag{4.12}$$

To simplify the presentation, it is assumed here that all end-nodes indexed by $i$ are connected to the small business $i$.

### 4.5.2  With Local Storage

The second scheme assumes that storage systems are installed at some small businesses and can be charged using solar power generated by the local PV system. However, excess solar production cannot be shared with other loads, even in the same balancing zone. Each controller aims at (1) charging the EV at the maximum rate, (2) minimizing curtailment of solar power produced by the local PV system, and (3) minimizing the use of conventional power from the grid, subject to the constraint that there is no export of real power to the grid. Thus, the controller implements the following rules in the given order:

$$p_i^e(t) = \overline{p}_i^e(t) \tag{4.13}$$

$$p_i^s(t) = \min\{\overline{p}_i^s(t), p_i^l(t) + p_i^e(t) + \underline{p}_i^b(t)\} \tag{4.14}$$

$$p_i^b(t) = \min\{\overline{p}_i^b(t), p_i^l(t) + p_i^e(t) - p_i^s(t)\} \tag{4.15}$$

As before, all end-nodes indexed by $i$ are connected to the small business $i$.

In the next chapter, we evaluate the performance of our control scheme, comparing it with these two benchmarks.

## 4.6  Chapter Summary

Many utilities in Europe and North America are experiencing the effects of high penetration of distributed PV systems and EVs on their radial distribution systems. Future distribution systems are anticipated to accommodate even higher penetrations of these technologies, threatening service reliability, impairing power quality, and reducing the efficiency of these systems under existing planning and operation paradigms. The synergy between EV chargers and PV inverters can be used to cancel out their effects on distribution feeders and simultaneously achieve the objectives defined by the utilities. An optimal control framework is proposed from which a decentralized scheme is derived to control EV chargers, PV inverters, and storage systems that are connected to low-voltage distribution networks. The proposed control is myopic, relies on end-node measurements, and requires a model of the distribution network. It enables power sharing within the balancing zones and is designed to address potential voltage, reverse flow, and congestion problems in distribution systems. The next chapter evaluates the efficiency and feasibility of the proposed control through power flow analysis.

## Reference

1. Katiraei F, Sun C, Enayati B (2015) No inverter left behind: Protection, controls, and testing for high penetrations of PV inverters on distribution systems. IEEE Power Energy Mag 13(2):43–49

# Chapter 5
# Evaluation

**Abstract** This chapter introduces an extensible, platform independent, smart grid simulation framework that combines discrete event and power flow simulation building blocks with AMPL, an optimization environment allowing the use of many commercial solvers. Extensive simulations are then performed using this framework to confirm that the proposed control scheme satisfies the operating constraints of the distribution system, and compare its efficiency with the two benchmark schemes presented in the previous chapter.

## 5.1 Simulation Framework

Evaluating the mechanisms devised for the control of a vast number of active end-nodes connected to low-voltage distribution feeders requires a simulation framework that supports

(a) creating large-scale simulation scenarios, each corresponding to a particular realization of several stochastic processes,
(b) jointly simulating the models developed for different aspects of the grid and the communication between measurement devices, active end-nodes, and controllers, considering its latency,
(c) solving various optimization problems formulated for the grid, and
(d) performing (multi-phase) power flow analysis.

Several commercial and open source software packages have been developed to perform each of these functions; however, no existing software fully supports the features required for running a large-scale smart grid simulation [5]. It is also not straightforward to piece together off-the-shelf simulators since they are not necessarily built for the same platform or have compatible input and output formats. Moreover, it is necessary to update the parameters of the power system simulator based on optimal control decisions computed by an external optimizer. This calls for the design of an extensible, platform independent simulation framework that couples the existing simulators and numerical computing environments, facilitates the exchange of intermediate results between these packages, and provides a unified

O. Ardakanian et al., *Integration of Renewable Generation and Elastic Loads into Distribution Grids*, SpringerBriefs in Electrical and Computer Engineering, DOI 10.1007/978-3-319-39984-3_5

API for defining various simulation scenarios. This section presents such a simulation framework. This framework was originally developed to validate the decentralized control scheme proposed in Chap. 4; however, it can be used to assess other schemes aiming to optimize and control distribution grids.

### 5.1.1   Architecture

Figure 5.1 depicts the simulation framework developed in this work, which combines a simulation tool for power distribution systems, called OpenDSS [7], with AMPL® [2], a powerful optimization environment. The simulation engine, developed in MATLAB® [10], coordinates the execution of these two software systems and provides a simple API which enables the users to load a test distribution network, upload or generate synthetic EV mobility and renewable energy traces, create models for loads and active end-nodes, define a control scheme, run discrete-time simulations, and collect performance results.

Figure 5.2 shows data and control flow between different parts of this simulation framework. An arrow depicts data/control flow, and a box represents a MATLAB

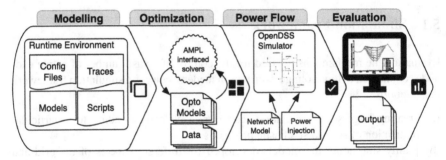

**Fig. 5.1**  Architecture of the simulation framework

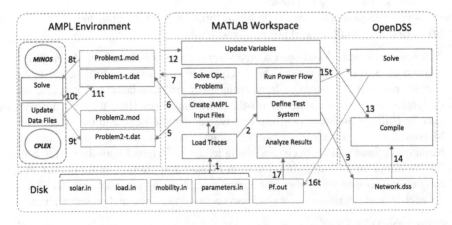

**Fig. 5.2**  Data and control flows between components of the simulation framework

function, a simple script that initiates some operation in AMPL or OpenDSS, or a data file stored on the disk. The arrows are numbered according to their execution order and those that must be executed in every simulation time slot are labeled with a number followed by 't'. Note that this architecture allows OpenDSS, AMPL solvers, and the simulation engine to run on different platforms communicating over TCP sockets.

## 5.1.2 Interactions Between Software Components

Consider a simulation run that involves solving a sequence of optimization problems in every time slot to determine the optimal control of a certain population of active end-nodes connected to a power distribution system. The MATLAB simulator builds the network model, describing the topology of the distribution system and points of connection of loads and active end-nodes, and can generate load, solar irradiance, and EV mobility traces based on some stochastic models. Once the models are loaded and traces are generated, the simulator creates the `network.dss` file which contains an OpenDSS-compilable network model, and several data files of the form `problemX-t.dat`, each containing AMPL parameters for an optimization problem, X, in a time slot, t.

In the next step, AMPL is called to load the optimization problems from `problemX.mod` files and assign values to the optimization parameters using data provided in `problemX-t.dat`. An AMPL-interfaced solver is then invoked to solve an optimization problem in a given time slot and append the optimal solution to the data file pertaining to the next optimization problem of this time slot, if there is any.

Once the optimization problems are solved for a time slot, the obtained solutions are used to update the data files corresponding to the next time slot. Finally, the MATLAB simulator updates the `network.dss` file with the optimal solutions, i.e., the control decisions, and calls the `compile` script in OpenDSS. The power flow simulator then compiles the distribution network model defined in `network.dss`, runs the power flow simulation for every time slot, and stores bus voltages and branch flows in `pf.out`. The simulation engine is notified when power flow simulations end to corroborate the feasibility of the control decisions and perform other post-processing steps. This concludes the simulation run.

## 5.1.3 Programming Interface

The following functions have been implemented in MATLAB to run a simulation scenario and analyze the results:

**Load Traces** Read data files from the disk. These files include load profiles for residential and commercial customers, and randomly generated solar and EV mobility traces. Note that these traces can either be generated by the simulator using some models or come from existing data sets.

**Define Test System** Build the distribution network model and create `network.dss` which will be compiled by OpenDSS.

**Create AMPL Input Files** Create AMPL data files, each describing parameters of an optimization problem.

**Solve Optimization Problems** Invoke the solver suitable for solving the optimization problem in a given time slot.

**Update Variables** Fetch optimal control decisions found by AMPL, update variables in the MATLAB environment, and modify the `network.dss` file.

**Run Power Flow** Call OpenDSS to preform power flow calculations for every time slot and write back the complex bus voltages and branch flows.

**Analyze Results** Examine voltage profiles and branch flows and write the final results to the disk.

The Bash scripts defined in the AMPL environment are as follows:

**Solve** Load the optimization problem model from `problemX.mod`, initialize its parameters by reading the data provided in `problemX-t.dat`, configure the selected solver, solve the optimization problem, and store the optimal solution.

**Update Data Files** Append the solution to the data file corresponding to the next time slot.

Finally, the OpenDSS COM interface provides these methods:

**Compile** Compile the distribution network model described in `network.dss`.

**Solve** Run power flow analysis to compute bus voltages and power flows given real and reactive power injected or consumed at each bus.

## 5.2   Simulation Scenarios

This section describes simulation scenarios that are used in the rest of this chapter. A scenario describes the number of inelastic loads and active end-nodes that are connected to each bus of a given distribution network. To obtain concrete simulation results, it is necessary to make numerous assumptions about the distribution system. We have tried our best to be as realistic as possible in the choice of these simulation parameters, recognizing that the results may change if these parameters are modified. Nevertheless, conclusions and insights gained from these simulations are relatively insensitive to the actual parameter choices.

The results of extensive simulations are presented in Sect. 5.3. For every penetration level of active end-nodes, 7 simulations are performed using traces obtained for 7 days in the summer (i.e., one simulation per day), and the average and the standard deviation of the parameters of interest are computed across these runs. The length of each time slot is set to 1 min in all simulations.

### 5.2.1 Test Distribution System

The proposed control scheme is evaluated on a variant of the IEEE 13-bus test feeder [9], which is a three-phase unbalanced radial distribution system supplied by a 5 MVA substation transformer stepping down the voltage from 115 to 4.16 kV. This radial system is modified as explained below. Recall that the proposed control scheme relies on a model that considers loads and active end-nodes connected to each phase separately, ignoring the coupling between phases. However, to understand how far this approach can be pushed, we evaluate it in Sect. 5.3 in a three-phase system through power flow simulations that take into account the coupling between phases.

Figure 5.3 shows primary distribution feeders and buses that comprise this radial system. A load bus represents a transformer connection where a distribution transformer supplies a low-voltage distribution network and downstream household and business loads. It is assumed that each low-voltage distribution network constitutes a balancing zone, depicted by dashed boxes in Fig. 5.3. Hence, distribution

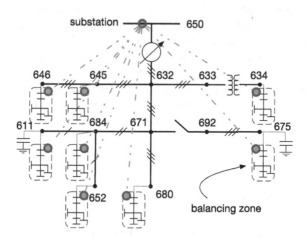

**Fig. 5.3** The one-line diagram of our radial test system, where slashes across each line indicate the number of phases. Balancing zones are depicted by dashed boxes connected to selected load buses. A low-voltage distribution network within a balancing zone is connected to each load bus. A communication network that forms a logical tree (*dotted lines*) over the distribution system connects the substation controller to balancing zone controllers, depicted by *circles*, and also to end-nodes (not represented here)

transformers are installed at the edge of balancing zones. This implies that real power cannot be injected into the network at load buses (but reverse flows within the balancing zones may be permitted, depending on the nature of the control scheme). Due to the lack of a realistic model for low-voltage distribution networks, it is assumed that demands of end-nodes within a balancing zone are aggregated at the corresponding load bus. Nevertheless, this approach can be extended to study the entire distribution grid if the low-voltage distribution network model is available.

In these scenarios the switch between buses 671 and 692 is closed, and shunt capacitors connected to buses 675 and 611 are switched on at all times. It is assumed that loads are single-phase connected between a phase and neutral. The single-phase power flow model discussed in Sect. 3.3 is incorporated into the optimization problems; hence, the coupling between phases is ignored in the computation of optimal controls (though not in the simulations). To simplify the model of the test system that is used in these optimization problems, the following assumptions are made: (a) the voltage magnitude at bus 650 is fixed at 4.16 kV, (b) the substation voltage regulator tap setting is fixed and known, and (c) the 500 kVA three-phase transformer between buses 633 and 634 is replaced with three 167 kVA single-phase transformers. Note that power flow simulations are performed on the standard test system *without* making these simplifying assumptions. The setpoint associated with a line is 90 % of its ampacity at 50 °C and the setpoint associated with a transformer is 90 % of its rated capacity. Following current practices, bus voltage magnitudes are required to stay within ±5 % of the nominal distribution voltage.

We adopt a plausible layout of loads at the distribution level, which can be viewed as educated guesses. Table 5.1 specifies how inelastic loads, including homes and businesses, are connected to the test system, noting that the figures provided in this table are per phase and node, and EV chargers, PV inverters, and storage systems are installed only at small businesses. We considered scenarios with 100, 200, 300, 400, and 500 PV systems and the same number of storage systems which are distributed in the distribution system according to Table 5.1.

A communication network connects the measurement devices installed at households and businesses with active end-nodes to the controller of the corresponding balancing zone, and also balancing zone controllers to the substation controller as depicted in Fig. 5.3.

### 5.2.2   Load Profiles

The test distribution network supplies a total of 3300 households and small businesses connected to selected buses as described in Table 5.1. It is assumed that demands of households and small businesses are approximately the same so we treat them interchangeably. To evaluate the decentralized control algorithm and examine its impacts on fast timescale dynamics of the grid, high-frequency electricity demand data of a large number of households are required. However, this data set is not publicly available owing to regulations that prevent utilities from sharing fine-grained

**Table 5.1** Loading condition of simulation scenarios

| Bus | 680 | | | 634 | | | 675 | | | 645 | | 646 | | 684 | | 652 | 611 |
|---|---|---|---|---|---|---|---|---|---|---|---|---|---|---|---|---|---|
| Phase | a | b | c | a | b | c | a | b | c | b | c | b | c | a | c | a | c |
| No. inelastic loads | 450 | 450 | 450 | 50 | 50 | 50 | 300 | 300 | 300 | 50 | 50 | 200 | 200 | 50 | 50 | 150 | 150 |
| No. EV chargers | 20 | 20 | 20 | 10 | 10 | 10 | 10 | 10 | 10 | 10 | 10 | 10 | 10 | 10 | 10 | 10 | 10 |
| Pct. storage | 10% | 10% | 10% | 5% | 5% | 5% | 5% | 5% | 5% | 5% | 5% | 5% | 5% | 5% | 5% | 5% | 5% |
| Pct. PV systems | 10% | 10% | 10% | 5% | 5% | 5% | 5% | 5% | 5% | 5% | 5% | 5% | 5% | 5% | 5% | 5% | 5% |

Simulation scenarios differ in the total number of PV systems and the total number of storage systems that are connected to the network current measurements from individual premises. Therefore, synthetic load profiles are used for the purpose of simulation. In particular, the load profiles are generated using the Markov models developed in [3] for household electricity consumption during on-peak, mid-peak, and off-peak periods. These models are derived from fine-grained measurements of electricity consumption in 20 Ontario homes over four months. A 1-min time resolution is chosen for synthesizing the load profiles to match the time resolution of our solar traces.

Using identical Markov models to generate the load profile of all homes and businesses results in a relatively smooth substation load during on-peak, mid-peak, and off-peak periods. However, the load abruptly changes at period boundaries. To avoid these abrupt transitions, we modulate the mean power consumption levels of our reference Markov models in every time slot such that the aggregate load at the substation resembles the Ontario demand in the first seven days of July 2014, which is shown in Fig. 5.4. Specifically, a correction factor is computed for every time slot by comparing the sum of all load profiles with the Ontario demand. The consumption level of all loads in every time slot is then multiplied by the correction factor computed for that time slot. This eliminates the abrupt changes.

The reactive power consumption of every home or business is assumed to be 30 % of its real power consumption in every time slot. This corresponds to a power factor of about 0.95 at the loads, which is typical for residential loads. Power flow calculations indicate that the peak demand of inelastic loads at the substation is 4.50 and 4.23 MW with and without losses, respectively. Thus, distribution losses amount to approximately 6 % of the demand. It also shows that the substation transformer is not congested over the simulation interval, in the absence of the active end-nodes.

**Fig. 5.4** The Ontario demand (5-min resolution) in the first week of July 2014

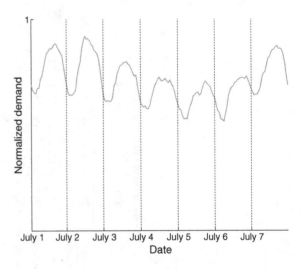

**Fig. 5.5** Solar irradiance data from a measurement site in Southern Great Plains

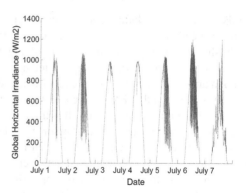

### 5.2.3 Solar Traces

Lacking high resolution solar data from Ontario, we obtained one-minute solar irradiance data for the week of July 1–July 7, 2003 from the Southern Great Plains atmospheric radiation measurement site in north-central Oklahoma [12], as shown in Fig. 5.5. This data set is used as a reference to generate solar traces for the installed panels. Specifically, the reference is scaled up such that the peak available power of a single PV installation is uniformly distributed in the range 4–5 kW, which is reasonable for a rooftop solar system, while the rated apparent power capacity of PV inverters is set to 5 kVA. Simulations are carried out for 100, 200, 300, 400, and 500 panels which are distributed in the network as described in Table 5.1.

### 5.2.4 Storage

Storage systems are assumed to be installed at every small business with a PV installation so that the excess solar generation can be stored locally. Hence, the number of storage systems is always equal to the number of PV systems in every simulation scenario. Table 5.1 describes the distribution of storage systems in the test system. The maximum charge and discharge powers of storage systems are set to 10 kW, their capacity is set to 5 kWh, and their charge and discharge efficiencies are assumed to be 95 %. At the beginning of every simulation, the SOC of all storage systems is assumed to be zero.

### 5.2.5 EV Model

Table 5.1 describes how 200 Level 2 EV chargers are connected to load buses. It is assumed that EV chargers are only installed at businesses (and not at homes) that also have PV installations and storage systems.[1] A maximum of one EV is connected to each charger at a time, a Level 2 charger imposes a maximum load of 7.2 kW on the distribution network, the capacity of an EV battery is 24 kWh (i.e., the capacity of a Nissan Leaf [11]), its charge efficiency is 95 %, and the SOC of all EVs is 0.5 upon arrival. Hence, the initial energy demand of every EV is 12 kWh.

It is assumed that EVs arrive and connect to the chargers (located at businesses) every day starting from 8am, following a Poisson distribution with parameter $\mu = \frac{200}{90} = 2.2$ arrivals per minute. Poisson arrivals have also been used by other work in the literature [4]. It is also assumed that EVs disconnect from the chargers 8 h after their arrival. Thus, the number of active chargers varies over time starting from 0 at 8am, rising to the full number by approximately 9:30am and starting to decline at 4pm, reducing to 0 by approximately 5:30pm.

## 5.3  Results

To examine the effects of uncontrolled EV charging and solar generation, power flow simulations are performed on the modified IEEE 13-bus test system in Sect. 5.3.1 and Sect. 5.3.2, respectively. Distribution network problems that arise in these cases motivate the design of a scheme that jointly controls EV chargers, PV inverters, and storage systems to achieve the utility-defined objectives, while addressing these problems. Section 5.3.3 validates the feasibility of the proposed control through power flow analysis, and quantifies the benefits of sharing and using storage within a balancing zone by comparing the efficiency of the decentralized control scheme with the two benchmarks defined in the previous chapter.

The integrated simulation framework presented in Sect. 5.1 is used to perform simulations for the specific scenarios that are described earlier. It is worth noting that all simulations are preformed on a dedicated optimization server with a 12-core processor and 500GB of memory, and CPLEX® and MINOS solvers are employed to solve the linear and nonlinear convex optimization problems described in Sect. 4.3. The OpenDSS simulator is configured to automatically control tap settings in power flow simulations to limit voltage fluctuations, as would be the case in normal operation, unless otherwise stated.

### 5.3.1  The Effect of Uncontrolled EV Charging

We first consider the case where only 200 EV chargers (and no PV or storage systems) are installed at small businesses, as described in Table 5.1. Hence, the distribution

---

[1]The only exception is the scenario in which there are 100 PV panels and 100 storage systems; hence, PV panels are fewer than EV chargers. In this scenario, the other 100 EV chargers are installed at randomly selected businesses.

substation serves 3300 homes and businesses and 200 chargers in this case. Two uncontrolled EV charging scenarios are studied here; the first scenario assumes that all chargers are Level 1 (a maximum load of 1.8 kW per charger) and the second one assumes that all of them are Level 2 (a maximum load of 7.2 kW per charger). In both scenarios chargers start charging at their maximum rate upon arrival of vehicles.

We run power flow analysis for both scenarios to obtain branch flows and bus voltages. Figure 5.6 shows the effect of uncontrolled EV charging on the substation transformer loading in both scenarios in the first day of our simulation. It can be seen that uncontrolled charging of EVs in the Level 2 charging scenario overloads the substation transformer once most chargers become active. Should Level 1 charging be adopted, the transformer loading does not exceed its nameplate rating. However, Level 1 charging extends the average charging time from 100 to 400 min (which is still acceptable). Expectedly, uncontrolled charging of EVs does not result in reverse power flow in both scenarios.

We now focus on the impact of uncontrolled charging on voltage profiles. One aspect we need to take into account is that voltage drop along the feeder can be remedied by careful choice of the voltage regulator tap position. To study this, we compare the voltage drop when tap positions can and cannot be changed every minute. First consider the case where regulator taps cannot be controlled at a fast timescale. In this case, the taps need to be fixed to a position at which bus voltages will remain within the tolerance limits over the simulation interval, when the grid only supplies household and business demands. We specifically use the tap setting +8, neutral, and +10 (each step is 0.625 %) for phase a, b, and c, respectively, which is a plausible setting for this loading condition. Table 5.2 shows the minimum and maximum voltage levels recorded in our power flow simulations for both scenarios. It can be seen that uncontrolled Level 2 EV charging results in undervoltage at bus 646. Should tap operations be permitted as fast as once per minute (or several minutes) to restore load voltage to normal, simultaneous charging of the entire EV population

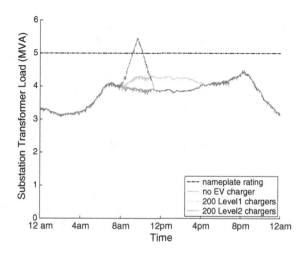

**Fig. 5.6** The effect of uncontrolled Level 1 and Level 2 EV charging on the substation transformer load

**Table 5.2**  Voltage magnitudes (p.u.) per phase and node for uncontrolled EV charging scenarios without optimizing voltage regulator tap settings

|       | 200 L1 chargers | | 200 L2 chargers | |
| ----- | ----- | ----- | ----- | ----- |
|       | Max   | Min   | Max   | Min   |
| 650a  | 1.000 | 1.000 | 1.000 | 1.000 |
| 650b  | 1.000 | 1.000 | 1.000 | 1.000 |
| 650c  | 1.000 | 1.000 | 1.000 | 1.000 |
| 632a  | 1.038 | 1.025 | 1.038 | 1.021 |
| 632b  | 0.985 | 0.969 | 0.985 | 0.968 |
| 632c  | 1.046 | 1.025 | 1.046 | 1.015 |
| 671a  | 1.025 | 0.998 | 1.025 | 0.992 |
| 671b  | 0.983 | 0.959 | 0.983 | 0.959 |
| 671c  | 1.036 | 0.998 | 1.036 | 0.980 |
| 680a  | 1.020 | 0.990 | 1.020 | 0.982 |
| 680b  | 0.980 | 0.953 | 0.980 | 0.953 |
| 680c  | 1.031 | 0.990 | 1.031 | 0.971 |
| 633a  | 1.038 | 1.024 | 1.038 | 1.019 |
| 633b  | 0.984 | 0.968 | 0.984 | 0.967 |
| 633c  | 1.046 | 1.024 | 1.046 | 1.013 |
| 634a  | 1.033 | 1.018 | 1.033 | 1.008 |
| 634b  | 0.979 | 0.962 | 0.979 | 0.956 |
| 634c  | 1.042 | 1.018 | 1.042 | 1.002 |
| 692a  | 1.025 | 0.998 | 1.025 | 0.992 |
| 692b  | 0.983 | 0.959 | 0.983 | 0.959 |
| 692c  | 1.036 | 0.998 | 1.036 | 0.980 |
| 675a  | 1.023 | 0.995 | 1.023 | 0.989 |
| 675b  | 0.982 | 0.956 | 0.982 | 0.956 |
| 675c  | 1.035 | 0.996 | 1.035 | 0.977 |
| 645b  | 0.978 | 0.959 | 0.978 | 0.955 |
| 645c  | 1.042 | 1.019 | 1.042 | 1.007 |
| 646b  | 0.974 | 0.954 | 0.974 | **0.949** |
| 646c  | 1.040 | 1.015 | 1.040 | 1.004 |
| 684a  | 1.023 | 0.995 | 1.023 | 0.988 |
| 684c  | 1.034 | 0.995 | 1.034 | 0.974 |
| 652a  | 1.018 | 0.988 | 1.018 | 0.979 |
| 611c  | 1.033 | 0.993 | 1.033 | 0.971 |

Voltage limit violations are printed in boldface

would not result in any voltage problem in both scenarios. Nevertheless, this would cause excessive wear on the voltage regulator, which translates into higher operation and maintenance costs, and is therefore not desirable [1, 14].

### 5.3.2  The Effect of Uncontrolled Solar Generation

We now study the case where a certain population of PV panels is installed at small businesses, as described in Table 5.1. We assume that no storage or EV charger is installed in this network, PV inverters only produce real power, their real power output is not throttled by the operator, and excess solar generation can be transferred to loads in the same balancing zone. We gradually increase the number of PV installations from 0 to 500 (0–15 % penetration) and perform power flow studies for each case. Note that, in some jurisdictions, even in 2014, a penetration rate of 20 % has already been achieved [13].

We first focus on the impact of uncontrolled solar generation on voltage profiles. Similar to the case of uncontrolled EV charging, we assume that regulator taps cannot be controlled on a fast timescale. We fix the taps using the same setting described in Sect. 5.3.1. Table 5.3 shows the minimum and maximum voltage levels recorded in our power flow simulations. It can be seen that overvoltage occurs at several buses, such as 634 and 645, when the number of PV installations exceeds 400. Should tap operations be permitted as fast as once per minute, our studies show that voltage does not increase beyond the permissible threshold at these penetration rates, even in the case of 500 PV systems. Again, recall that voltage regulators are meant to be controlled infrequently and this would cause excessive wear on them.

As we expected, distribution lines and transformers are not overloaded in these simulations because distributed solar generation reduces their load. Instead, reverse flow is observed at buses 634, 645, and 684 when the number of PV installations exceeds 200. Figure 5.7 shows the effect of uncontrolled solar generation on the substation transformer load for different penetration rates and the direction of power flow at bus 634 in the case of 500 PV systems. It can be seen that the net load decreases drastically during the day when solar power is available and ramps up again in the evening; this is widely known as the 'duck curve' [6]. Furthermore, it can be seen that real power flows from bus 634 towards bus 632 in most time slots when the sun is shining. This reverse flow can cause severe problems discussed in [8, 15]. Interestingly, most EVs are connected to chargers at small businesses in this time interval, suggesting that the synergy between EV chargers and PV inverters could enhance power system reliability and address network problems that are likely to occur at high EV and PV penetration rates. This motivates the design of the proposed control scheme.

### 5.3.3  Evaluating the Proposed Control

This section compares the decentralized control scheme with the two benchmark schemes defined in Sect. 4.5. Recall that in all three schemes, both EVs and PV systems are present. In the benchmark schemes, only local observations are used to make control decisions, whereas in our scheme, a central controller coordinates

**Table 5.3** Voltage magnitudes (p.u.) per phase and node for uncontrolled solar generation scenarios without optimizing voltage regulator tap settings

|       | 100 PVs–3 % | | 200 PVs–6 % | | 300 PVs–9 % | | 400 PVs–12 % | | 500 PVs–15 % | |
|-------|-------|-------|-------|-------|-------|-------|-------|-------|-------|-------|
|       | Max   | Min   | Max   | Min   | Max   | Min   | Max   | Min   | Max   | Min   |
| 650a  | 1.000 | 1.000 | 1.000 | 1.000 | 1.000 | 1.000 | 1.000 | 1.000 | 1.000 | 1.000 |
| 650b  | 1.000 | 1.000 | 1.000 | 1.000 | 1.000 | 1.000 | 1.000 | 1.000 | 1.000 | 1.000 |
| 650c  | 1.000 | 1.000 | 1.000 | 1.000 | 1.000 | 1.000 | 1.000 | 1.000 | 1.000 | 1.000 |
| 632a  | 1.038 | 1.025 | 1.038 | 1.025 | 1.038 | 1.025 | 1.040 | 1.025 | 1.041 | 1.025 |
| 632b  | 0.985 | 0.969 | 0.985 | 0.969 | 0.985 | 0.969 | 0.987 | 0.969 | 0.989 | 0.969 |
| 632c  | 1.046 | 1.025 | 1.046 | 1.025 | 1.049 | 1.025 | **1.053** | 1.025 | **1.056** | 1.025 |
| 671a  | 1.025 | 0.998 | 1.025 | 0.998 | 1.025 | 0.999 | 1.026 | 0.999 | 1.028 | 0.999 |
| 671b  | 0.983 | 0.959 | 0.983 | 0.959 | 0.983 | 0.959 | 0.983 | 0.959 | 0.984 | 0.959 |
| 671c  | 1.036 | 0.998 | 1.036 | 0.998 | 1.041 | 0.998 | 1.047 | 0.998 | **1.053** | 0.998 |
| 680a  | 1.020 | 0.990 | 1.020 | 0.990 | 1.020 | 0.990 | 1.021 | 0.990 | 1.023 | 0.990 |
| 680b  | 0.980 | 0.953 | 0.980 | 0.953 | 0.980 | 0.953 | 0.980 | 0.953 | 0.981 | 0.953 |
| 680c  | 1.031 | 0.990 | 1.031 | 0.990 | 1.036 | 0.990 | 1.043 | 0.990 | 1.049 | 0.990 |
| 633a  | 1.038 | 1.024 | 1.038 | 1.024 | 1.038 | 1.024 | 1.040 | 1.024 | 1.041 | 1.024 |
| 633b  | 0.984 | 0.968 | 0.984 | 0.968 | 0.985 | 0.968 | 0.987 | 0.968 | 0.989 | 0.968 |
| 633c  | 1.046 | 1.024 | 1.046 | 1.024 | 1.049 | 1.024 | **1.053** | 1.024 | **1.056** | 1.024 |
| 634a  | 1.033 | 1.018 | 1.034 | 1.018 | 1.037 | 1.018 | 1.039 | 1.018 | 1.042 | 1.018 |
| 634b  | 0.979 | 0.962 | 0.981 | 0.962 | 0.984 | 0.962 | 0.987 | 0.962 | 0.990 | 0.962 |
| 634c  | 1.042 | 1.018 | 1.043 | 1.018 | 1.048 | 1.018 | **1.053** | 1.018 | **1.058** | 1.018 |
| 692a  | 1.025 | 0.998 | 1.025 | 0.998 | 1.025 | 0.999 | 1.026 | 0.999 | 1.028 | 0.999 |
| 692b  | 0.983 | 0.959 | 0.983 | 0.959 | 0.983 | 0.959 | 0.983 | 0.959 | 0.984 | 0.959 |
| 692c  | 1.036 | 0.998 | 1.036 | 0.998 | 1.041 | 0.998 | 1.047 | 0.998 | **1.053** | 0.998 |
| 675a  | 1.023 | 0.996 | 1.023 | 0.996 | 1.023 | 0.996 | 1.025 | 0.996 | 1.027 | 0.996 |
| 675b  | 0.982 | 0.956 | 0.982 | 0.956 | 0.982 | 0.956 | 0.982 | 0.956 | 0.983 | 0.956 |
| 675c  | 1.035 | 0.996 | 1.035 | 0.996 | 1.040 | 0.996 | 1.046 | 0.996 | **1.052** | 0.996 |
| 645b  | 0.978 | 0.959 | 0.978 | 0.959 | 0.980 | 0.959 | 0.983 | 0.959 | 0.986 | 0.959 |
| 645c  | 1.042 | 1.019 | 1.042 | 1.019 | 1.046 | 1.019 | **1.051** | 1.019 | **1.053** | 1.019 |
| 646b  | 0.974 | 0.954 | 0.974 | 0.954 | 0.977 | 0.954 | 0.980 | 0.954 | 0.983 | 0.954 |
| 646c  | 1.040 | 1.015 | 1.040 | 1.015 | 1.044 | 1.015 | 1.048 | 1.015 | **1.052** | 1.015 |
| 684a  | 1.023 | 0.995 | 1.023 | 0.995 | 1.023 | 0.996 | 1.025 | 0.996 | 1.027 | 0.996 |
| 684c  | 1.034 | 0.995 | 1.034 | 0.995 | 1.040 | 0.995 | 1.047 | 0.995 | **1.054** | 0.995 |
| 652a  | 1.018 | 0.988 | 1.018 | 0.988 | 1.019 | 0.988 | 1.022 | 0.988 | 1.025 | 0.988 |
| 611c  | 1.033 | 0.993 | 1.033 | 0.993 | 1.040 | 0.993 | 1.047 | 0.993 | **1.054** | 0.993 |

Voltage limit violations are printed in boldface

decisions. Moreover, in the benchmark schemes, PV panels are not allowed to inject power into the balancing zone, but in our scheme, power sharing is allowed within each balancing zone. Figure 5.8 shows the total available solar power and the total real power output of PV inverters for different control schemes, when 100 PV panels

**Fig. 5.7** The effect of solar generation with uncontrolled inverters on the substation transformer load for different PV penetration rates for a typical day

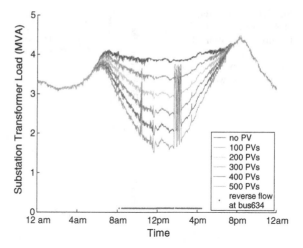

**Fig. 5.8** Total power output of PV inverters over a day for different control schemes in the case that 100 PV panels are deployed in the distribution network

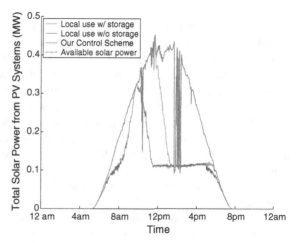

and 100 storage systems are connected to the test distribution network. It can be seen that with the decentralized control it is possible to use all of the solar energy available in every time slot. This is because excess solar generation can always be stored or consumed by loads that are in the same balancing zone at this PV penetration rate. Observe, also, that the two benchmark schemes use much less solar power due to curtailment, especially when storage is unavailable.

When the number of PV installations (and the number of storage systems) increases to 400, the decentralized scheme would have to curtail solar generation in some time slots to prevent reverse flow and maintain voltage within the bounds.[2]

---

[2]We attribute abrupt changes in the total real power output of PV inverters when our control is implemented to changes in the number of active chargers, load fluctuations, reverse flow restrictions, and storage capacity constraints.

**Fig. 5.9** Total power output of PV inverters over a day for different control schemes in the case that 400 PV panels are deployed in the distribution network

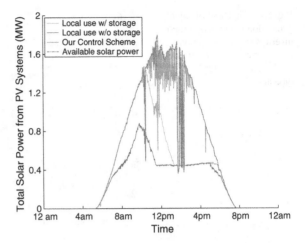

This can be seen in Fig. 5.9, where after all the EVs have been charged, no more than 0.4 MW of solar capacity can be used. In contrast, our control results in much less curtailment compared to the other two schemes as shown in Fig. 5.10. The same observation is made when the number of PV installations increases to 500.

We now compare the performance of the proposed control scheme in two cases, where storage is available and unavailable, with the two benchmark schemes. Figure 5.10 shows that the average amount of solar energy curtailed by different schemes over the period of a day. The proposed control does not result in solar curtailment when there are fewer than 300 PV/storage systems, or when there are fewer than 200 PV systems (but no storage). Even when the number of PV installations increases to 400 and 500, respectively, the proposed control results in, on average, 90.9 % and 78.3 % less curtailment compared to the first local control scheme, and 85.3 % and 65.7 % less curtailment compared to the second local control scheme. Furthermore, simulation results suggest that sharing is more effective in reducing curtailment than using even 5 kWh storage per PV location. This result is very insightful for electric utilities in that sharing is cost-free unlike expensive storage systems.

Control schemes can also be compared in terms of their use of conventional power supplied by the grid. Figure 5.11 shows that the proposed control scheme reduces the use of conventional power (by up to 5 %) by displacing it with solar power. Note that the studied control schemes are similar in terms of the energy supplied to EVs since all of them manage to fully charge EVs before they leave in our simulation scenarios. However, if the EVs left earlier (for example after 6 h instead of 8 h), the proposed scheme would allocate power to connected EVs in a proportionally fair manner, while benchmark schemes would not provide fairness and could result in starvation.

**Fig. 5.10** Average solar
energy curtailed by different
control schemes over the
period of a day (lower is
better). Error bars represent
one standard error

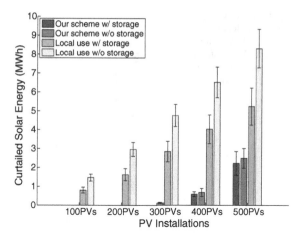

**Fig. 5.11** Average use of
conventional energy by
different control schemes
over the period of a day
(lower is better). Error bars
represent one standard error

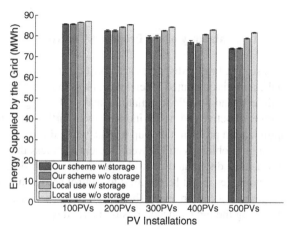

Finally, power flow studies find that the proposed open-loop controller does not
cause *any* voltage, congestion, or reverse flow problem in all simulation scenarios.
As an example of the resulting operation, Fig. 5.12 shows the substation loading over
the first day of our simulation. It can be seen that our control successfully prevents
overloads, confirming that using setpoints that are 10 % below the nameplate ratings
is sufficient to compensate for inaccuracies of the simplified DistFlow model.[3] We
caution that, we did see that higher equipment setpoints, e.g., setting them equal to
the nameplate ratings and 5 % below the nameplate ratings, often led to infeasible
optimization problems. Thus, the results are sensitive to the choice of using 90 % of
the rated equipment capacity as the setpoint.

---

[3]Electric utilities have a rough estimate of resistive losses in their distribution circuits, enabling
them to appropriately choose the equipment setpoints.

**Fig. 5.12** Substation transformer loading over a day for 200 EV chargers and different PV and storage penetration rates using the proposed control

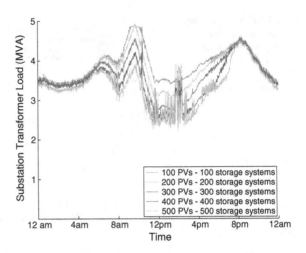

#### 5.3.3.1   Scalability of the Proposed Control Scheme

This section briefly discusses scalability of the proposed control scheme. The optimal control is computed efficiently (in less than 1 s) on a single machine in scenarios with less than 400 PV installations and 400 storage systems. However, should the number of active end-nodes increase even further, solving the first (substation level) optimization problem takes up to 30 s, while the second (balancing zone level) optimization problem is still solved efficiently, in less than a few seconds. This implies that a fully distributed control scheme in which decision making is delegated to the end-nodes might be required to control active end-nodes when their penetration increases to a certain level. The design of this scheme is quite complex and requires exploiting the hidden decomposition structure of the optimization problems.

### 5.4   Chapter Summary

Assessing the advanced optimization and control schemes developed for power systems requires an integrated simulation framework capable of performing power flow analysis, solving the underlying optimization problems in a decentralized manner, and simulating arrivals and departures of EVs and the communications between measurement devices, controlled nodes, and controllers. A powerful smart grid simulator that performs all these functions is introduced in this chapter and is used later to validate the decentralized control scheme developed in Chap. 4. It is shown through extensive simulations and power flow studies on a radial test system that this scheme successfully addresses voltage, reverse flow, and congestion problems, allocates available power in a proportionally fair manner among active EV chargers, harnesses as much solar energy as possible using storage and sharing, and minimizes the use of conventional power. Moreover, simulation results corroborate that the decentralized control scheme, which stores and shares solar generation within balancing zones, significantly reduces solar curtailment compared to the benchmark schemes.

# References

1. Agalgaonkar Y, Pal B, Jabr R (2014) Distribution voltage control considering the impact of PV generation on tap changers and autonomous regulators. IEEE Trans Power Syst 29(1):182–192
2. AMPL Optimization (2016) (retrieved) AMPL. http://ampl.com/products/ampl
3. Ardakanian O, Keshav S, Rosenberg C (2011) Markovian models for home electricity consumption. In: Proceedings of the 2nd ACM SIGCOMM workshop on green networking, pp 31–36
4. Bae S, Kwasinski A (2012) Spatial and temporal model of electric vehicle charging demand. IEEE Trans Smart Grid 3(1):394–403
5. Büscher M, Claassen A, Kube M, Lehnhoff S, Piech K, Rohjans S, Scherfke S, Steinbrink C, Velasquez J, Tempez F, Bouzid Y (2014) Integrated Smart Grid simulations for generic automation architectures with RT-LAB and mosaik. In: IEEE smart grid communications, pp 194–199
6. California ISO (2016) (retrieved) Fast Facts. http://www.caiso.com/Documents/FlexibleResourcesHelpRenewables_FastFacts.pdf
7. EPRI (2016) (retrieved) Simulation Tool – OpenDSS. http://smartgrid.epri.com/SimulationTool.aspx
8. Katiraei F, Agüero J (2011) Solar PV integration challenges. IEEE Power Energy Mag 9(3): 62–71
9. Kersting W (2001) Radial distribution test feeders. In: IEEE PES winter meeting, vol 2, pp 908–912
10. MathWorks (2016) (retrieved) MATLAB. http://www.mathworks.com/products/matlab/
11. Nissan (2015) (retrieved) Nissan Leaf. http://www.nissan.ca/en/electric-cars/leaf/versions-specs/version.s.html
12. NREL (2016) (retrieved) Atmospheric radiation measurement program. http://www.nrel.gov/midc/arm_rcs/
13. Parkinson G (2015) (retrieved) Rooftop solar to cut total grid demand to zero in South Australia. http://reneweconomy.com.au/2015/rooftop-solar-to-cut-total-grid-demand-to-zero-in-south-australia-32943
14. Paudyal S, Canizares C, Bhattacharya K (2011) Optimal operation of distribution feeders in smart grids. IEEE Trans Ind Electron 58(10):4495–4503
15. Walling R, Saint R, Dugan R, Burke J, Kojovic L (2008) Summary of distributed resources impact on power delivery systems. IEEE Trans Power Deliv 23(3):1636–1644

# Chapter 6
# Conclusion

**Abstract** The focus of this brief has been on a control framework for active end-nodes to mitigate emerging operational and technical problems and fulfill various environmental, societal, and business objectives. This chapter summarizes the achieved goals, highlights the limitations of this framework, and suggests several avenues for future work.

## 6.1 Summary of Achieved Goals

This brief investigated the challenges of integrating variable-power DER, such as renewable energy systems and storage technologies, and high-power elastic loads, such as plug-in electric vehicles, into low-voltage distribution grids and the pivotal role of information and communications technology (ICT) in overcoming these challenges. Leveraging low-cost broadband communications and pervasive monitoring of the end-nodes in distribution networks, a control framework is proposed to solve multi-objective multi-constraint control problems in quasi real-time. A decentralized control mechanism capable of simultaneously achieving various user-level and system-level objectives is then developed as an extension of the mechanisms that are currently in place for balancing the grid. This is a nontrivial task as these objectives are often competing.

In particular, the synergy between solar PV generation and EV chargers is used to tackle distribution system problems, increasing the degree of penetration of both PV systems and EVs that can be reliably accommodated in existing power systems. The active end-nodes are controlled using a decentralized scheme that solves linearized power flow equations using real-time measurements of the demand of inelastic loads and the state of the active end-nodes. Since the underlying optimization problems are convex, the optimal control can be found both quickly and efficiently. This control scheme is fair to active EV chargers, and permits sharing of solar power and the use of storage systems within a balancing zone, thereby reducing solar curtailment and the use of conventional power from the grid. It has been shown using numerical simulations, which are based on realistic load and solar generation traces and stochastic EV arrivals and departures, that this control outperforms the schemes that limit solar

© The Author(s) 2016
O. Ardakanian et al., *Integration of Renewable Generation and Elastic Loads into Distribution Grids*, SpringerBriefs in Electrical and Computer Engineering, DOI 10.1007/978-3-319-39984-3_6

generation to the local demand in terms of solar curtailment and conventional energy use. The power flow analysis confirmed that the proposed control does not cause overloads, overvoltage and undervoltage conditions, or reverse flows as long as the setpoints are selected judiciously.

The control framework proposed in this work is inspired by the design of well-established resource allocation and flow control algorithms that have been originally developed for the Internet. For example, the notion of proportional fairness often used in scheduling problems has been extended to the EV charging problem.

## 6.2  Limitations and Future Work

This section presents existing challenges and limitations of the proposed control framework that could be addressed in future work.

### 6.2.1  TCP-Style Control for Active End-Nodes

In Chap. 4, exploiting real-time measurements of the demand of inelastic loads and the state of active end-nodes, two optimization problems are solved to obtain the optimal control in every iteration. Unlike the second optimization problem which is decomposed and solved independently for each balancing zone, the first optimization problem cannot be solved in a fully distributed fashion. This is because some constraints, such as the simplified DistFlow equations, couple the active end-nodes connected to different balancing zones. Thus, the proposed algorithm for solving the first problem does not scale with the size of the distribution network and the number of active end-nodes. Specifically, we have seen that solving the first optimization problem takes several seconds once the number of PV systems and the number of storage systems exceed 400.

What is needed is a TCP-like feedback control mechanism for active end-nodes. A potential solution, similar to what has been done in [1], would be to decompose the centralized optimization problem, which relates real and reactive power injections to bus voltages, into several subproblems, each solved independently by an active end-node. The decoupled problems are coordinated by a master problem using Lagrangian multipliers [3]. This makes it possible to develop a simple feedback control mechanism for active end-nodes based on in-network rather than end-node measurements. Control would be fully distributed and is therefore more scalable compared to the proposed open-loop control mechanism. Designing this control mechanism is complicated because there are many coupled variables and coupling constraints.

### 6.2.2 Generalizing to Unbalanced Multi-Phase Distribution Systems

The power flow equations are solved separately for each phase of the distribution network to obtain optimal control decisions. However, distribution networks are usually unbalanced and ignoring the coupling between different phases introduces some error into our analysis. A possible direction for future work is to substitute this model with a distribution power flow model for unbalanced multi-phase networks, similar to the linear approximation proposed in [2] or the generic distribution power flow model proposed in [4].

Note that loads are typically modelled as voltage dependent components in distribution systems. To simplify power flow calculations, a constant complex power load model is used in Chap. 4. A better load model is also a fruitful avenue for future work.

### 6.2.3 Optimizing Capacitor Banks and Load Tap Switching Operations

Conventional distribution system operation has been chiefly concerned about voltage and reactive power control using local measurements with distribution loss minimization being the operational objective in most cases. This is generally achieved by solving a distribution optimal power flow problem to control operations of transformer LTCs and switched capacitors [4].

Recall that the optimization problems solved in Chap. 4 to compute optimal control of active end-nodes also involves power flow calculations for the distribution system. This indicates the possibility of incorporating transformer LTCs and switched capacitors into our control problem to jointly optimize operations of EV chargers, solar PV inverters, storage systems, and switching of taps and capacitor banks. A similar approach has been taken in [5] to control EV chargers and taps and capacitor switching decisions. The main challenge here is that active end-nodes, and LTCs and capacitors must be controlled on two different timescales; thus, combining them into a single control problem requires careful consideration of the control timescales.

## 6.3 Concluding Remarks

The increasing penetration of elastic loads and distributed renewable generation, along with the introduction of measurement, communication, and control technologies in power distribution systems has several implications. Specifically, pervasive measurement and communication increases interactions between customers, system operators, and independent producers, providing new opportunities to improve reliability, as well as cost and carbon efficiency of the grid. Additionally, the integration of active end-nodes into low-voltage residential distribution networks enables the introduction of several new environmental, societal, and business objectives for which the grid has not been designed originally. Control, especially in the last mile of the distribution network, plays a key role in accomplishing these goals. However, existing grid controls are incapable of solving multi-objective multi-constraint problems that involve a large number of active end-nodes and new control solutions have not been defined yet to achieve recently introduced objectives of the active end-nodes. This work attempted to fill this gap in the literature by developing a decentralized algorithm for the control of active end-nodes in quasi real-time.

Despite the novelty of this approach, it has certain limitations. Firstly, the proposed control scheme could result in suboptimal or infeasible control decisions in unbalanced, three-phase radial and mesh distribution systems. Secondly, it relies on power flow calculations to obtain a feasible control, and therefore, assumes the knowledge of the system admittance matrix, which might not be available in some cases. Lastly, the substation controller is still a bottleneck, limiting the scalability of the control algorithm. These limitations present ample opportunities for future work.

# References

1. Ardakanian O, Keshav S, Rosenberg C (2014) Real-time distributed control for smart electric vehicle chargers: from a static to a dynamic study. IEEE Trans Smart Grid 5(5):2295–2305
2. Gan L (2015) Distributed load control in multiphase radial networks. PhD thesis, California Institute of Technology
3. Palomar D, Chiang M (2006) A tutorial on decomposition methods for network utility maximization. IEEE J Sel Areas Commun 24(8):1439–1451
4. Paudyal S, Canizares C, Bhattacharya K (2011) Optimal operation of distribution feeders in smart grids. IEEE Trans Ind Electron 58(10):4495–4503
5. Sharma I, Canizares C, Bhattacharya K (2014) Smart charging of PEVs penetrating into residential distribution systems. IEEE Trans Smart Grid 5(3):1196–1209

# Index

© The Author(s) 2016
O. Ardakanian et al., *Integration of Renewable Generation and Elastic Loads
into Distribution Grids*, SpringerBriefs in Electrical and Computer Engineering,
DOI 10.1007/978-3-319-39984-3

Printed in the United States
By Bookmasters